Estimating and Cost Control in Electrical Construction Design

WILLIAM C. MILLER, P.E.

A Construction Publishing Company Book

VNR VAN NOSTRAND REINHOLD COMPANY
NEW YORK   CINCINNATI   ATLANTA   DALLAS   SAN FRANCISCO
LONDON   TORONTO   MELBOURNE

Van Nostrand Reinhold Company Regional Offices:
New York   Cincinnati   Atlanta   Dallas   San Francisco

Van Nostrand Reinhold Company International Offices:
London   Toronto   Melbourne

Copyright © 1978 by Litton Educational Publishing, Inc.

Library of Congress Catalog Card Number: 77-20846
ISBN: 0-442-12203-9

All rights reserved. No part of this work covered by the copyright hereon may be reproduced or used in any form or by any means—graphic, electronic, or mechanical, including photocopying, recording, taping, or information storage and retrieval systems—without permission of the publisher.

Manufactured in the United States of America

Published by Van Nostrand Reinhold Company
450 West 33rd Street, New York, N.Y. 10001

Published simultaneously in Canada by Van Nostrand Reinhold Ltd.

15 14 13 12 11 10 9 8 7 6 5 4 3 2 1

Library of Congress Cataloging in Publication Data

Miller, William Charles, 1927–
  Estimating and cost control in electrical construction design.

  "A Construction Publishing Company book."
  Includes index.
  1. Electric engineering—Estimates. I. Title.
TK435.M54   621.319'24   77-20846
ISBN 0-442-12203-9

Preface

This book is intended to be used primarily by design engineers and architects or students of the building arts. Its main objectives are to enable the user to establish electrical construction costs early in the design process and to present methods of controlling those costs.

WILLIAM C. MILLER

Introduction

Two major constraints common to all construction projects are time and money. A good design engineer will work within these constraints to develop construction documents that can produce buildings within the budget and on schedule. This book should help the designer to reach those goals. Methods of construction cost control with values based on good estimating procedures are discussed.

The book deals with the various types of and methods used in preparing electrical estimates. Cost models are explained and illustrated, as are their worth as part of the cost control system. Charts have been devised for use early in design work to illustrate the variation in costs with equipment selection.

It is recognized that the major cost decisions are made early in the design process when the least amount of material for gauging costs is available. The charts should help provide yardsticks in these areas. Cost-control systems involve the owner as part of the picture and illustrate how to present project cost changes in relation to cost targets.

Not every design engineer is, excusably, an estimator. Few college courses are devoted to such non technical, mundane topics. But the engineer will surely be held responsible if his phase of a project misses its cost target. Aside from a reluctance to deal in pricing, engineers have a real obstacle to overcome in producing accurate estimates. This stems from the fact that vendors are more apt to deal seriously with and quote accurately to their point of sale: the contractor. Admittedly, the task of the contractor's estimator is easier. All he has to do is a counting and pricing exercise. Decisions affecting these costs are made in the design office, which has considerably less contact with market conditions. Nevertheless, estimate we must, and so in this book we have placed estimating tools and yardsticks into the hands of those people who establish the key cost components of a project.

Everyone has his own funny (sad) story of costs going berserk. However, times are changing, and overruns in time and money are no longer acceptable. Even the Federal Government has initiated what it calls design to cost programs for defense procurement. Contracts for design services are now being written limiting cost overrides, and the limits are being enforced. The engineering office producing a set of documents that are bid by contractors over estimated price may have a serious redesign job on its hands. We are stressing good estimating and cost control by the head end of the design team.

Estimating methods devised for this book allow the estimator, regardless of his background, to follow the cost effect of his design choices. We feel that merely listing costs of labor and material denies understanding to people who are not electrical estimators. Each of the electrical parts of a building will be described in sufficient detail to illustrate its part in the cost picture. Our intention is not to instruct electrical engineers or contractors' estimators in electrical design, but to put the proper estimating tool into the hands of those who can use it to greatest advantage.

Establishing a value for electrical work early in a building program need not be a blind, hopeful stab in the dark. Electrical work, like work in any of the other trades, proceeds from a rational design. There are certain essential elements and an interrelationship of the parts permitting a logical system of estimating to be established. Experience with large numbers of electrical designs for the several building types has enabled these relationships to be quantitified herein and represented in the charts and tables.

Some sections deal with components of estimates, estimate types, and methods. Value engineering is dealt with briefly, and one section discusses life-cycle analysis. There are also descriptions of the items contained in the estimation of each part of a building electrical system. Charts show costs of components. Reporting forms for estimates, cost models, and life-cycle analyses have been included.

Sections dealing with the actual estimating are all indexed alike. A simple line diagram sets out a system of numbers for the components which are used throughout subsequent mention of these items. The reporting form, cost model, descriptive text, and cost charts are all keyed alike.

The book attempts to be concise. Descriptive material accompanying the cost charts is short, presupposing familiarity on the part of the user. More detail is provided in cost-control sections, since this is the type of material the author believes is needed in the design office.

Contents

Preface	iii
Introduction	v
1. Construction Costs	**1**
2. Cost Control	**32**
3. Material and Labor	**67**
4. Building Components	**88**
5. Charts	**115**
6. Sample Estimates	**153**
Appendix	**165**
Index	**183**

1. Construction Costs

SUMMARY

Estimating is part of our everyday activity. We all know how far away a place might be, how much something weighs, and sometimes even what worth can be put on a piece of furniture, an automobile, or electrical equipment. These estimates serve as gauging devices, ways of putting some order into random situations so that we can deal with the data. If we know how far away our destination is we can plan travel time; knowing the weight of an object might save a back or a piece of lifting equipment; the worth tells us if we can afford the equipment being priced.

The more detailed and comprehensive an estimate the greater its value. While detail might be limited by available information, other components not so limited must be fully utilized. This chapter contains a description of these components. Information available is used at different times in the design process to produce different periodic estimates. There are, of course, several methods of estimating electrical work and these are illustrated.

Accurate estimating is the key ingredient to several other important activities. The information gathered for the estimate will be used in life-cycle analyses and various cost-control operations. From total costs, we can backtrack into manpower requirements and into scheduling assistance.

There is also a cost for accurate estimating, which should not be left without consideration. This and the results of poor estimating on the design fee are another part of this chapter.

ESTIMATE COMPONENTS

An estimate is an opinion. The estimation of the cost of electrical construction is the subject of this book. Basically, an estimate consists of two major blocks: the costs related to the physical nature of the installation, such as lighting, feeders, panels, etc.; and the costs a contractor adds for the operation of his business. The costs involved in equipment and its installation are shown in Chapters 4 and 5. The second class of costs, or job-related costs, will be discussed here.

Job-Related Costs

Job-related costs roughly break down into costs dependent on field conditions and business overhead items. Field conditions, broadly grouped, are the so-called general conditions. When a construction project is bid through a single general contractor, much of general-condition work is a part of his job. However, where work is bid as separate prime contracts by trade, each of the trades may be responsible for some part of the work that would have been part of the general construction contract. It is up to the design team to delineate clearly the responsibility of each prime contractor so as to make sure that all general-condition work is covered, but covered only once. The design engineer, perhaps, should have a checklist on which functions are assigned to each contract. Where an item, for example bonding or insurance, requires partial contribution by each trade to represent the total, a tally mark will appear next to each prime contractor.

A typical list of those items affecting the electrical contract may read as follows:

1. Performance bond
2. Liability insurance
3. Temporary light and power—energy cost
4. Watchmen
5. Surveys
6. Permits and fees
7. Field-office trailer
8. Storage equipment
9. Field-office supplies
10. Drawing reproduction
11. Coordination drawings
12. CPM input
13. Hand tools
14. Equipment rental
15. Excavation and backfill
16. Cutting and patching
17. Demolition
18. Protection of trees or existing structures
19. Shoring, planking, fencing, protection
20. OSHA safety requirements
21. Project telephone, intercom, public address
22. Material handling equipment at the site
23. Rubbish removal and cleanup
24. Temporary water
25. Temporary sanitary facilities
26. Temporary heat and air conditioning
27. Snow removal
28. Temporary weather protection
29. Project sign
30. Record drawings
31. Project manager
32. Period of construction

Of course, these items do not all necessarily apply to every job, while some projects may have a longer list. Nor are all the items equal in cost. We will consider only the major expense items, but leave the entire list as a guide for use in tightening up specification of general conditions.

General Conditions

Whether the contract is a prime contract or a subcontract will determine part of the jobsite overhead. Certain work is required of a prime contractor which is not performed by subcontractors. Primes are most closely involved in project coordination.

The subcontractors may be required to provide their own temporary facilities, drainage, snow removal, cleanup, etc. This item will include the contractor's normal site-expense items—office trailer, storage lockers, equipment, hand tools, telephone, blueprinting. Complex jobs for public agencies may require an office staff at the site for processing the paper generated. Somebody will be involved in updating drawings, coordinating shop drawings with the other trades, attending CPM meetings, processing change orders and field directives, estimating work in place for monthly requisitions, and taking care of as-built drawings. All of this is an overhead cost applied to the job as a factor. The usual range of these costs is from 4% to 10% of the direct cost of construction.

Performance Bond

This bond guarantees completion of work required under the contract. This item may or may not be in the electrical contract, depending on whether or not there are single or separate prime contractors. This is a bottom-line type of item amounting to about one-half of 1% of the contract price. From the contractor's viewpoint it is a cost written into the final dollar value of the signed contract.

Liability Insurance

Insurance should be required of all contractors to protect against public liability and property damage. This guards the contractors against injury to others by their operations at the project. Costs vary in relation to exposure. An injury to a person can occur on any project, but obviously the risk is greater where more workers are involved. Chances of injury to outsiders or property are greater in crowded urban locations than in a more open situation.

Watchmen

Depending on how the contract is written the electrical contractor may be responsible for the whole or part of security services. On some projects this can be fairly extensive, requiring fences and formal patrols by one or several men. Cost should be estimated for the project life and the proper percentage assigned to this contract. Some projects will require security television, lighting, and other systems.

Temporary Facilities

Providing a job office and storage facility can lead to substantial expenditures. Where a project requires technical and clerical help, a reasonable work place must be provided. This usually appears in the form of a trailer with air conditioning and sanitary facilities. Utility connections must be made and maintained. Equipment may include typewriters, calculators, filing equipment, reproduction equipment, drafting supplies, surveying materials. All of this has to be physically maintained, and the whole kit has to be cleaned on a regular basis. The field telephone bill is also part of the field office cost. On a project of large physical size radio telephones may be based here to maintain contact with field crews. In addition to other temporary energy costs (such as heat and electricity) applied to the entire project, the electrical contractor may have to provide this energy to his field offices. On a large project of considerable duration these costs can be substantial.

The contractor must bear a share of the cost of the other temporary utilities assigned to him by the design team. A major item here can be the energy cost of electricity consumed by temporary facilities for the life of the job. This can range from temporary heat in a partially completed building to simply temporary light or nothing at all, the energy costs being borne by the owner. General conditions should be quite specific in assigning this item. Usual missing items are the energy costs for job trailers of all trades, temporary connections to operate electrical equipment associated with temporary heat when the heat itself is not being furnished by electricity, connections and energy for testing substantial equipment, hours of operation of the system, or its parts.

On projects to be built in isolated locations the electrical contractor may have to be assigned the cost of furnishing and installing a generator plant and electric distribution equipment. This would supply construction needs for electricity until permanent service is available. Fuel costs for this type of operation can be expensive and should be clearly assigned.

Where temporary heat is involved it will vary in its sophistication. There may be a 24-hour requirement or a day-night temperature requirement or a starting date not consistent with the start of cold weather but rather with project status. Whether the electrical energy involved in the heating system is the heating medium itself or merely for operating equipment required for temporary heat should be stated in the HVAC design section.

Hours of operation and loading can then be assigned costs based on local utility company rates. Incidental costs of temporary facilities should also be discussed with the utility company, insofar as connection charges are concerned.

Project Manager

A *Project manager* is used here as a collective noun. We mean to include the entire field staff assigned directly to a particular job and not working with mechanic's tools. This may include a job superintendent, engineer, draftsman, file clerks, or any combination of these facilities. This cost will always apply to the project, whether in large or small measure.

Cutting, Patching, and Demolition

These items strictly speaking should be part of the estimate rather than a job-related cost. They are listed here more or less as a notice that they are real costs and should be accounted for. In the case of an alteration, considerable work in relocating existing equipment may be involved. Finishing materials may be destroyed in the process and will have to be restored. These items are heavy in labor percentage. A fair estimate should be made of this work. All too frequently, designers not familiar with all the lost effort involved in this kind of production will minimize or overlook these items entirely.

Work of this type may be required to be performed at night or on weekends where the areas are occupied and operating. Temporary connections and bypassing may be required. Existing facilities may have to be maintained. In certain areas, physical access may be limited both as to manpower and material. All these impediments to normal work production will raise the cost. This work should clearly be provided for as part of contract documents, and should be estimated with generous amounts of labor.

Variable Factors

Only infrequently do design offices have the benefit of repeated use of a completed layout. This means that jobs are all different, and in order to relate them to each other job factors have to be used. In general these are items such as building type, working conditions, the nature and ability of the general contractor, and the bidding contractor's own familiarity with the type of work to be performed. In one list of these factors prepared by the National Electrical Contractors Association, there is a spread in cost from the lowest to the highest category of 100%. Then NECA suggests an additional 1% per floor. The four major categories are broken down into 22 subcategories. Then there are potential variable factors, such as nonproductive labor and labor productivity. Other items a contractor considers are:

1. Job supervision
2. Layout
3. Materials handling
4. Testing
5. Delays
6. Job size as it affects productivity
7. Change in work schedules
8. Engineer's approval delay
9. Vandalism
10. Administration
11. Engineering
12. Change-order processing
13. Building construction materials
14. Shape of building
15. Parallel or single conduit runs
16. Types and complexity of electrical systems
17. Density of equipment
18. Repetitive installations
19. Space conflicts with other trades
20. Equipment room space
21. Cooperation of the design engineer
22. Cooperation of the utility company
23. Cooperation of inspectors
24. Time schedule of job
25. Labor market
26. Material market
27. Money market
28. Nature of bidding competition

Of course it must be realized that this is only a general list. Each contractor has his own set of worry beads.

Fortunately for the design-office estimator, not many of these items affect prices in the preparation of his own estimates. Final prices taken at bid openings will all include these items in one form or another. Therefore they represent what in effect may be considered as part of the contractor's overhead. Any extraordinary features of the overall building design, site, labor market, or material market should be examined for potential effects on the project cost.

Should any particular factor in the list become outstanding in its departure from what might be considered normal, it should be evaluated separately.

Period of Construction

Project duration has a direct effect on project cost. This arises from the manner in which time affects crew size. A contractor likes to have the flexibility to permit his crew size to fluctuate in sensitive response to job progress. Short construction periods may require overtime work or large crew sizes.

Local union regulations and contracts will usually define the electrical crew composition. There are limits on the number of journeymen allowable under a working foreman. For example, if a foreman is permitted to work with tools until he has six journeymen, but over that becomes an administrator only, the contractor may see fit to limit his crew to the most productive combination. The ratio of apprentices to journeymen permitted will also be a factor in determining daily crew costs.

The electrical designer should feel obliged to examine the general project schedules to see that these will be realistic from the electrical contractor's point of view. In the event that CPM (Critical Path Method) scheduling is used, care should be taken in considering time requirements. Tasks can be broken down from the estimate line items into man-hours. This will give an indication of crew size. The schedule set by the CPM will order the events. Putting all this information together the designer can examine the job-manning question. Too short a construction period may require undue amounts of overtime or inefficient crew size. Too long a period may cause extra expense in administrative effort and the maintenance of temporary on-site facilities.

An examination of the typical manning curve in Fig. 1-1 will demonstrate some of the effects on cost of the length of the construction period. Early phases of the work go slowly, requiring a certain momentum to be developed. Clearing and grading with heavy equipment by the general contractor and little actual construction serves to keep the electrical work within very circumscribed bounds. With a normal construction period, the electrician can conveniently follow the general contractor as work areas become available. Eventually, as rough work nears completion and the finishing work begins, crew size is at a maximum. Crew size fluctuates as necessary to suit the job.

Under an accelerated schedule, problems develop. Available work areas are still limited at the outset, but become accessible in large blocks as the general contract work proceeds. It is necessary for the electrical contractor to increase crew size or work overtime. Either of these choices is expensive. In the example given, the reduction in time is approximately 30%. Assume a labor intensity factor of 0.70 (see Chapter 3); manpower will now cost between 15% and 30% more than the normal job. This comes about partly as the cost of overtime, partly because of incorrect crew size, and partly by loss of efficiency.

It should be noted that in the event a project requires a shortening of time schedule there will follow claims by the contractor for expenses of the type outlined above. If, on the other hand, the project time is extended (after letting the contracts), the contractors will

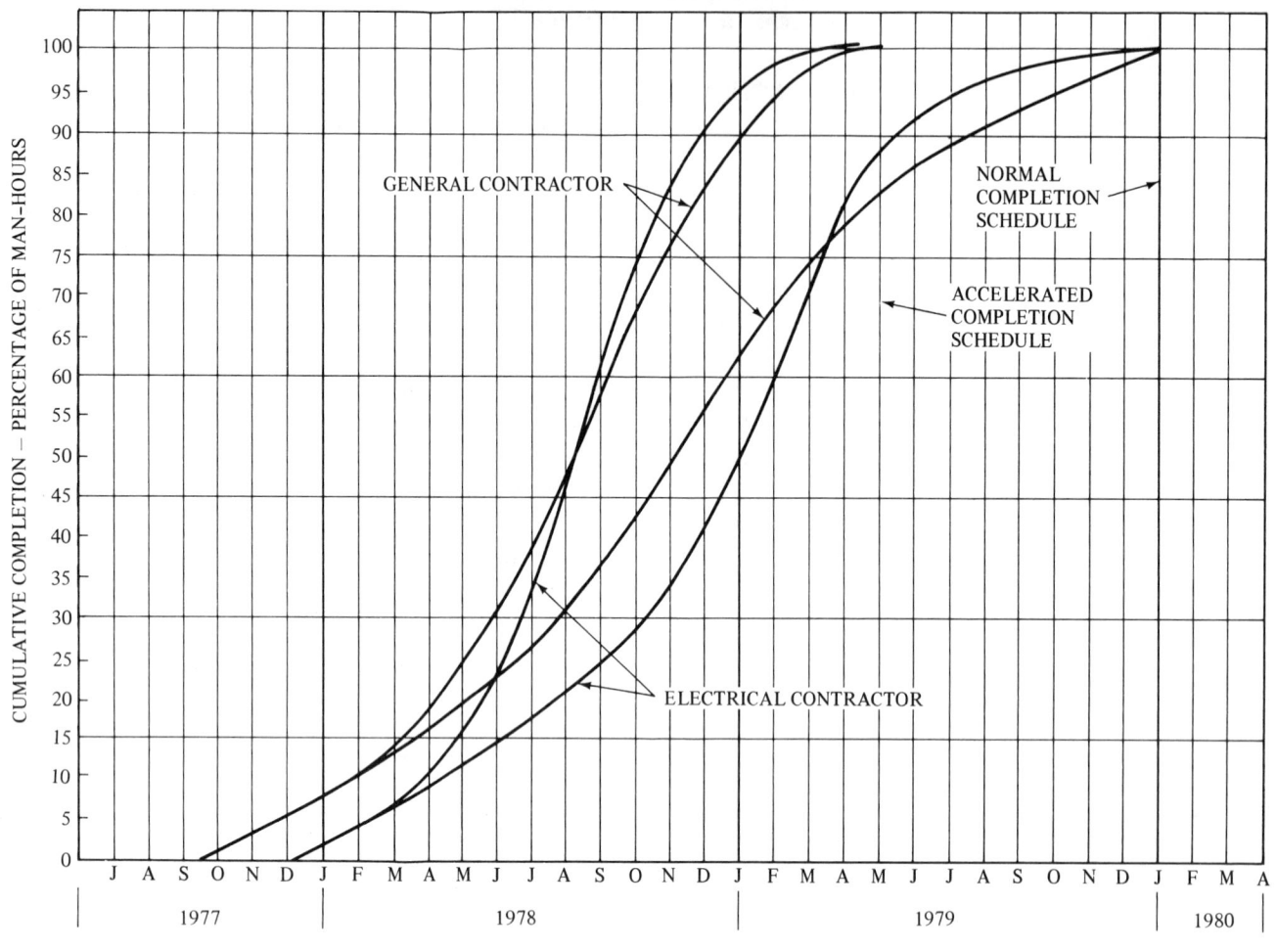

Fig. 1-1.

claim extended costs such as supervision, tool rental, home-office expenses, etc.

Selecting a reasonable time duration for the project is an important function of the designer. Any construction period other than what might be considered normal will result in increased costs.

Market-Related Costs

Contingencies

Generally speaking, there are usually two types of contingency considered in making an estimate. These are the estimator contingency and the design contingency.

The estimator contingency is effectively an ignorance factor where an outside estimator is involved. No estimator can so accurately read the designer's mind, in the early stages of design, as to see the full scope of the finished product. The estimator's experience allows him to gauge his relative ignorance of the final form and to simplify the designer's understatement. To an extent, allowance for this contingency also represents a protection for the estimator against being embarrasingly low too early in the design process. The theory is that adjustments of a design that raise the cost are always easier than the reverse. This type of contingency can easily be as high as 15%, reducing in later phases of the design work to zero. With in-house estimating or early estimating by the design engineer himself, this factor can be substantially reduced or eliminated.

The design contingency comes about as drawings are detailed toward completion, when the nuts and bolts are put in. There is always something new to be added, or somebody in the client's office has an idea, or there are other changes of this nature. These items have a tendency, left uncontrolled, to grow substantially. Methods of containing such price rises have been discussed under cost control.

Overhead

Direct costs applicable to any particular job that a contractor has underway can vary considerably. Some of the major considerations are the nature of the job, cooperation of the owner in rapid payment processing, and miscellaneous expenses involved in establishing the job. These can include such items as field office and expenses, engineering, change-order processing, and home-office support. Costs of this nature will vary radically from job to job, ranging between 3% and 15% of the direct job costs. A good, medium figure to use is 10%.

Profit

Methods of computing profit vary. We can call profit a discernable difference between the contractor's price and the definable costs. The amount allowed as a profit multiplier by the contractor when making a bid for a job will vary. Conditions of the market are a major factor. Lack of available projects and a large number of competing bidders will result in a shaving of profit as a bottom-line adder. To some extent the size of the project and the relative risk involved may also affect the contractor's estimate of a fair return. Generally, it is safe to say that profit will be what the contractor feels he can get. This will sometimes be evident from comparison of a good engineer's estimate with a quotation submitted by a contractor. For the purpose of establishing a working number to use until final, pre-bid conditions can be evaluated, a figure such as 10% is common.

Note that both overhead and profit (sometimes inseparable) are strongly determined by the job market at the time of bid. The estimator should consider these figures with an eye toward adjustment when final estimates are made.

Escalation

It is likely that either the price index or labor rate or both may change between the date of any particular kind of estimate and some time in the future. The difference between future price and present cost, divided by the present cost, may be defined as *escalation*. Historically, the trend of costs is upward, but under certain conditions costs may remain the same or even go down. Predicting these trends requires the use of various economic indicators, cost indices, and a crystal ball or two. Once determined, escalation is applied as a multiplying factor to the bottom-line figure of the estimate. An estimate should always bear a note indicating the point to which escalation has been taken. All price-line items are taken for the date of the estimate, and only the entire total is escalated. In this manner, one might notice both line-item price changes and escalation-factor changes as the time of drawing development proceeds through periods of market variations.

The accuracy of an escalation factor is usually higher over a short period than over a long one. Predicting the near future is somewhat more certain than the distant future. Actually, the most important date to be considered is the midpoint of construction, although other dates may be used for special purposes. In pricing out his job, the contractor attempts to anticipate labor and material costs over the period of actual construction time. This is when he will be taking deliveries and paying salaries. His average price over the project duration then will be at the approximate midpoint of the job. In his own estimate, the designer therefore has to aim at the same target date the contractor will use.

In order to improve total accuracy of the escalation factor, the midpoint of construction should be as short a distance in the future as possible. This can result from the shortening of any of the usual time periods involved; design, bidding, or construction. Another method, which is being used with increasing frequency, is the use of small-bid packaging to bring contractors on the job before the project may be completely designed or bid. The effect of this so-called fast track is to shorten the escalation period.

Example:

Figure 1-2 is a typical project master schedule. The key elements indicated are, design, bid and award, and construction. Figure 1-3 is an escalation chart for electrical work in the geographical region of the project. These two figures illustrate the use of escalation factors.

On the project master schedule (Fig. 1-2) we have indicated the midpoint of construction. Note that this is not necessarily the project midpoint, nor is it even likely to be. As was previously pointed out, it is the construction-phase timing that determines what market conditions will be faced by the contractor. It is therefore the construction period which must be considered in determining the escalating period. In some projects, it should be noted, the electrical construction midpoint will not coincide with the construction midpoint. If these

PROJECT MASTER SCHEDULE

PROJ OR W/PK	MILESTONES	1976	1977	1978	1979
1	MASTER SCHEDULE		S———————————|——————————————|——————X	
2	DESIGN		S————X		MIDPOINT OF CONSTRUCTION
3	BID & AWARD		S——X		
4	CONSTRUCTION			S———————————|——X	
5	INSPECTION & ACCEPTANCE				S—X

Fig. 1-2.

dates are expected to be widely different, due consideration should be given to determining the actual midpoint of electrical contract work, since that will be the date considered by bidders on electrical work.

Once the construction midpoint date is determined, it should be marked on the escalation curve (Fig. 1-3). The other significant date to be marked on the curve is the date on which the estimate was made. In the case of the illustration,

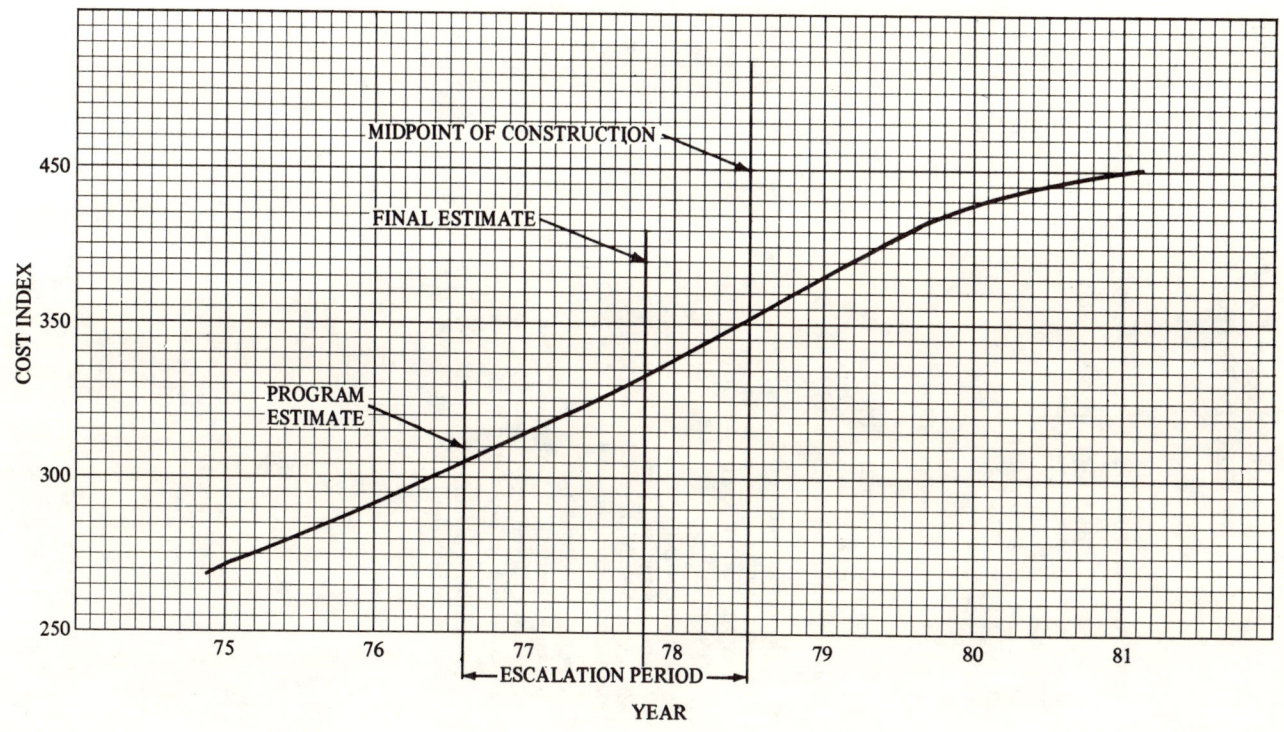

Fig. 1-3.

CONSTRUCTION COSTS 11

Building cost and price index roundup
Base: 1967 = 100

Name, Area and Type	1973 AV.	1974 AV.	1975 AV.	1975 Apr.	May	June	July	Aug.	Sept.	Oct.	Nov.	Dec.	1976 Jan.	Feb.	Mar.	Apr.	May	June
GENERAL PURPOSE COST INDEXES																		
ENR 20-cities; Construction Cost	177	185	206	—	201	205	209	211	211	213	213	213	214	215	216	217	217	223
ENR 20-cities; Building Cost	169	178	193	—	191	194	195	197	197	200	200	201	202	203	204	205	206	209
U.S. Commerce Department	152	173	190	138	189	190	190	189	190	191r	191r	193r	194r	194r	195r	—	—	—
BuRec. Denver, Bldgs.	155	175	196	193	—	—	197	—	—	200	—	—	201	—	—	202	—	—
Dodge Bldg. Cost	185	205	230	215	—	—	—	—	—	230	—	—	—	—	—	—	—	—
Factory Mutual Indus. Bldg.	163	176	200	—	—	—	202	—	—	—	—	211	211	—	—	—	—	—
Lee Saylor Inc.: Labor/Material	164	195	209	—	—	202	—	—	207	—	—	209	—	—	211	—	—	214
Means: Const. Cost	160	176	188	185	—	—	190	—	—	192	—	—	192	—	—	195	—	—
CONTRACTOR PRICE INDEXES—BUILDING																		
Austin: Central & Eastern U.S., Indus.	151	176	187	—	—	185	—	—	188	—	—	191	—	—	193	—	—	—
Fruin-Colnon: St. Louis, Industrial	160	183	202	193	200	198	199	207	210	210	210	211	214	217	217	—	—	—
Geo. A. Fuller	162	185	204	198	—	—	209	—	—	207	—	167	211	—	—	—	—	—
Lee Saylor Inc.: Subcontractor	162	188	167	—	—	183	—	—	175	—	—	—	—	—	172	—	—	171
Turner: General	163	190	198	198	—	—	199	—	—	198	—	—	199	—	—	201	—	—
Smith, Hinchman & Grylls: General#	169	176	186	183	184	186	187	189	189	189	191	191	192	193	194	196	196	—
H. F. Campbell: 17-cities. Mfg.	157	176	194	192	192	192	198	198	198	201	201	201	201	201	201	202	202	—
VALUATION INDEXES																		
American Appraisal: 30-cities. Indus.*	167	179	189	184	187	188	191	192	194	194	194	196	198	199	200	202	203	—
Boeckh index: 20-cities. Comml & Mfg.**	154	171	189	—	188	—	190	—	192	—	194	—	196	—	200	—	199	—
Marshall & Swift: Industrial	154	169	184	—	—	181	—	—	186	—	—	189	—	—	190	—	—	194
SPECIAL PURPOSE INDEXES																		
Nelson Refinery Cost: "Inflation" Index	163	182	200	196	197	199	200	204	204	206	206	207	208	—	—	—	—	—
Chemical Engineering plant cost	131	151	166	165	165	166	166	166	167	169	169	170	171	171	172	172	—	—
Port of New York Authority. Hangar Cost	157	175	194	191	191	192	196	196	196	199	199	198	199	199	200	199	—	—

*Adjusted for productivity. **reinforced concrete. #Smith, Hinchman & Grylls is an A-E FIRM.
Prepared by Boeckh—a division of American Appraisal Associates Inc.

Fig. 1-4.

two dates are shown. These indicate the date of the program estimate and the date of the final estimate. From the curve, cost indices are as follows:

a = Midpoint of construction cost index
b = Final estimate cost index
c = Program estimate cost index

Escalation factors are as follows:

$$\text{Program estimate factor (PEF)} = \frac{c}{a} = \frac{350}{305} = 1.148$$

$$\text{Final estimate factor} \quad \text{(FEF)} = \frac{c}{b} = \frac{350}{330} = 1.061$$

Assuming the program raw estimate was $1,000,000, the escalated cost at that time would have resulted in an estimated cost of

$$\$1,000,000 \times \text{PEF} = \$1,148,000$$

Now, assume that in the design phase, the usual program changes resulted in cost adjustments producing a final raw estimate of $1,050,000; then the final estimate would be

$$\$1,050,000 \times \text{FEF} = \$1,114,050$$

Presumably, the final estimate (which will be made on prices then current) will include any escalation that has already taken place between the two estimates. This is automatic, since both estimates use the market prices current at the time of the estimate.

In the illustration above, cost control has been excellent, maintaining the projected construction cost during the design development phase. As a measure of success, the designer should look back from the final estimate, to see if cost changes were within the actual market-index changes. In long or interrupted design periods, departures from the market index will sound the alarm of slipping cost control.

Various market indices are available. For example, in Fig. 1-4, an excerpt from *Engineering News Record,* some of the indices to be found are listed with past-year averages.

Example:

Another illustration might serve to point out the relative changes in escalation factor as the design process proceeds in time. Figure 1-5 projects a fixed annual escalation rate, rather than using one of the indices.

1. The schematic estimate should include escalation projected to the mid point of construction. As illustrated, this would amount to 9%.

2. The 50% design estimate should include escalation projected to the mid point of construction. In the illustration this is 6%.

3. The final estimate in the example should include an escalation of 3% to carry it to the mid point of the construction period.

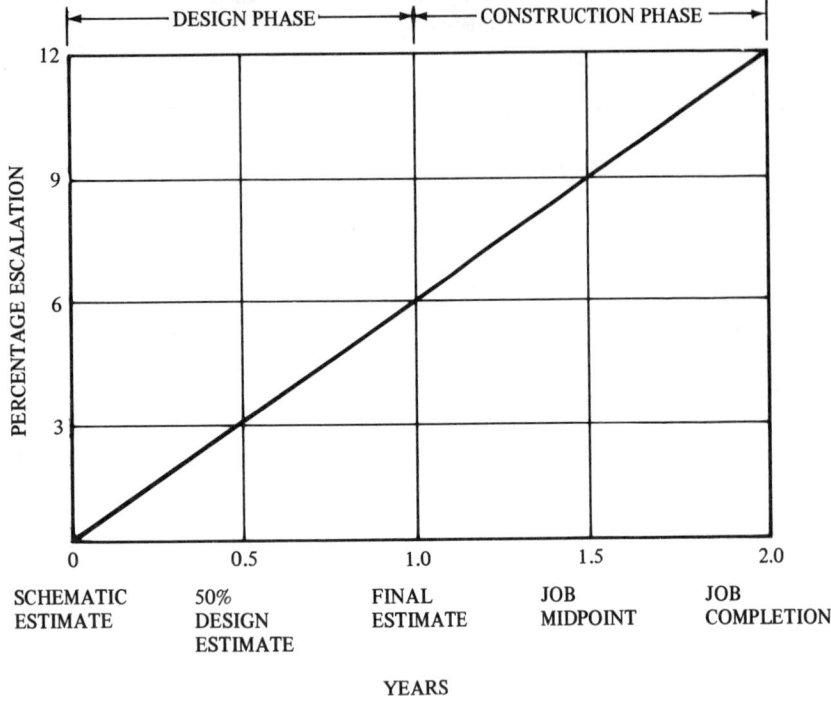

Fig. 1-5.

Market Conditions

The estimator must consider available manpower and materials for the project. Some projects will require special skills, others ordinary skilled workmen in a remote area where they may not be available. Construction market conditions in a particular location may be in some extreme condition, stressing the available labor, transportation, or materials market. Those factors that can be predicted should be assigned a value and applied to project cost. The bidding climate as determined by local construction activity will affect the bid price. When work is scarce and a large number of bidders can be attracted, bid prices tend to be lowered.

Conversely, in good times, bidders may be scarce. This results in a limited-market bid strategy, allowing the bidder a better chance of connecting with a job at a price he finds very favorable. This situation requires careful consideration of the proper time to put a job out for bid. If it is necessary to enter an unfavorable market, the designer should consider the inclusion of alternates as a means of using available funds without the cutback and rebid effort an overbid might otherwise require.

Number of Bidders—Adjustment Coefficients*
(4 Bidders = 1.00)

Number of bidders	Adjustment coefficient
1	1.349
2	1.134
3	1.046
4	1.000
5	0.962
6	0.946
7	0.929
8	0.915
9	0.903
10	0.890

If Bidders exceed 10, use coefficient for 10 Bidders.

*Adapted from a study published in *Engineering News Record*, no date.

Example:

The accompanying table relates bid results to the number of bidders. Given a project bid at 5,400,000 with 6 bidders, what would it have been if there were only 4 bidders?

$$5,400,000 \times \frac{1}{0.946} = \$5,708,345$$

Consider the effects of using the extremes in the bid group on the job illustrated:

with 10 bidders: bid price = $4,002,965

with 1 bidder: bid price = $6,067,415

These extremes are separated by 50% of the lower price!

Using the more common range of 4–6 bidders, as illustrated, results in a bid-price differential of 5.7%, which, incidentally, might be more than the design-engineering fee.

One of the points the example makes clear is that the timing of the bid period is quite important. Try not to go out to bid when some research shows several other jobs are taking up the interest of local contractors. Sometimes a postponement (or an acceleration) of the bid opening date to a more favorable time is a wise investment for the owner.

This bid timing, of course, requires the designer to do a certain amount of crystal-ball gazing. The author recognizes that long-range forecasting of any type is difficult. Forecasting the economic climate may be less accurate than forecasting the weather. One thing can be stated with some certainty; and bears repetition: short-range forecasts will be more accurate than long-range forecasts. Consequently, the designer assessing market conditions will do best with short periods

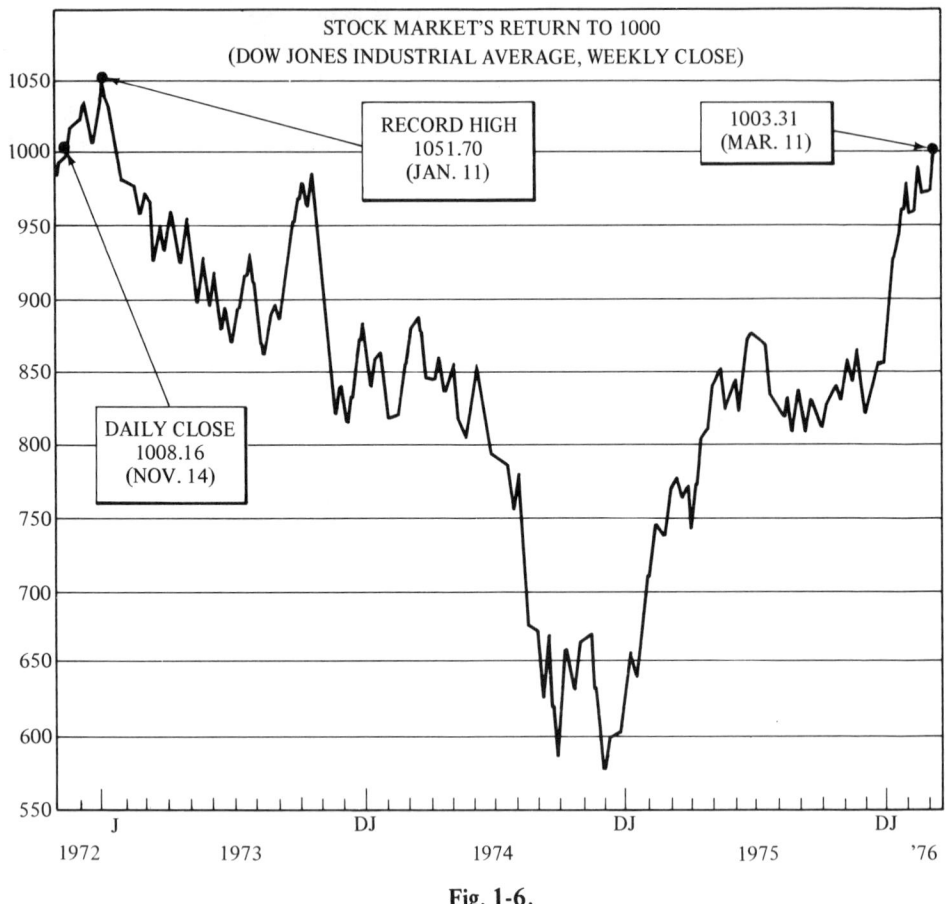

Fig. 1-6.

of construction and early bid dates. There is a margin of protection against error in escalation affecting the bid price. This lies in the fact that the bidders must guess, along with the designer, at labor and material escalation to the future mid point of construction. In this the designer and contractor are on more or less equal footing.

Labor and materials will escalate at different rates. Labor rates are somewhat more predictable, coming in the changes due to new contracts. These may run over a period of time. Materials, on the other hand, follow the market and economic climate.

Predicting these events with accuracy would indicate the designer might be better off in another more lucrative field. Figure 1-6, which is generally indicative of the material price index, shows the recent history of the Dow Jones Industrial Average for the New York Stock Exchange. Anybody who can really predict this curve's position on a particular date is wasting his time in engineering.

ESTIMATING METHODS

All estimates are basically as good as the available information. A set of machine design drawings or detailed final construction documents will be priced with very little cost spread by several experienced estimators. Obtaining a reasonable estimate of cost early in the design phase is a different story, since information is sketchy and quantities cannot be obtained. Still, estimates at this time are essential to guide the designer within certain cost ranges.

Cost-per-Square-Foot Estimate

Frequently, early estimates are made on a square-foot basis. An experienced estimator can look at a program and adjust standard square-foot prices to suit his feeling for costs of the project at hand.

The chart on page 180 is useful for this type of pricing. Such a chart can be used to provide a starting point in assembling a cost estimate. The specific costs listed and their adjustment upward or downward are very subjective matters. Charts of this type find great use as check figures. Costs per square foot will give an indication of scale with which one can begin to place electrical costs as a part of the total cost. The charts often have a fairly wide range of prices. These will account somewhat for variations in relative quality of the project. They should be used with some knowledge of the style of the architect: whether he tends to be elaborate or simple, uses innovative techniques or standard stock items, etc. The cost-per-square-foot figures, when used with judgement, are of great value in getting a first handle on what costs might be expected.

When using these figures, ascertain if they are based on net or gross square feet. Note considerations for subcontractor's overhead and profit as well as the same items for the general contractor. The figures will generally permit comparison within building types. They should not be used beyond the very earliest talking stages.

Other services supply bid-price reports with simple descriptions of the building's vital statistics. These are reported on a countrywide basis and may serve as a guide for comparison with a project at hand.

Gross features may be compared and adjusted upward or downward subjectively by the designer. (See page 167.)

If possible, several check systems should be used for comparison. Variations may serve to pinpoint errors. Agreement should provide greater confidence.

Cost-Capacity Estimates

A useful method of early estimating is based on the similarity in costs for similar projects. An office that specializes in designing a particular building type can accumulate its own data base for this kind of cost projection. In this system, buildings may be compared on an essential common element such as square feet, cubic feet, kilowatts, hospital beds, etc. Older projects are updated in cost by appropriate escalation factors obtained from one of the building indices and a projection is made to estimate the cost of the project under consideration on a current basis. The expression

$$C_b = C_a \left(\frac{S_b}{S_a}\right)^x$$

is used for the comparison, where

C_a, C_b = costs of project a and project b, respectively,
S_a, S_b = sizes of project a and project b, respectively, in units chosen to express the similarity (square feet, kilowatts, etc.),
x = cost-capacity factor, representative of the building type.

Example:

Find the estimate for electrical work for a public housing project scheduled for 400 apartments. Historical data show housing of 300 units at a cost of $10,500,000.00.
  Given:

C_a = $10,500,000, cost of 300 apartment project,
S_a = 300 apartments,
S_b = 400 apartments,
  x = 0.75 cost-capacity factor,

find

C_b = cost of 400 apartment project.

Solving the equation,

$$C_b = C_a \left(\frac{S_b}{S_a}\right)^x = 10{,}500{,}000 \left(\frac{400}{300}\right)^{0.75}$$
$$= \$13{,}028{,}000$$

If historical data has produced a cost capacity factor of 0.85 the result would be:

$$C_b = 10{,}500{,}000 \left(\frac{400}{300}\right)^{0.85}$$
$$= \$13{,}409{,}000.$$

Increasing the exponent cost-capacity factor by 13% has resulted in a cost increase of 2.9%. This type of estimate can safely be used only when cost information is from one's own historical base. Changes in design style from project to project would render the results nearly useless. In the example above the exponent was taken from a chart in William R. Park's book, *Cost Engineering Analysis* (see the bibliography). Most of the factors Park lists are for industrial-type buildings little subject to wide changes in design; in other words, the method is applied to a more utilitarian structure. The design of such buildings is governed by output of a product to a greater extent than is design of commercial or institutional buildings.

Some typical cost-capacity factors are given below:

Type of facility	Cost-Capacity factor	Units of capacity
Steam plant	0.75	pounds/hour
Electric generating plant	0.79	megawatts
Industrial building	0.67	square feet
Primary sewage-treatment plant	0.68	gallons/day

XYY ESTIMATORS INC.

Trade ELECTRICAL   Job: _____

Estimator _____   Project _____

Checker _____ App'd By _____ _____

Date _____ Revision _____ Type Est. FINAL _____

ITEM	CLASSIFICATION OF WORK	QUANTITY TOTAL	UNIT	RATE	AMOUNT	TOTAL COST
1	MAIN DISTRIBUTION AND PANELS					
	01 Distribution Panels - 480V	2	EA	1600		
	02 Panels - 480/277V	5		435		
	03 Panels - 120/208V	4		250		
	04 Disconnect Switches - 100A	2		70	8 790	
	05 24-Hour Time Switch	1		35		
	06 Automatic Transfer Switch - 100A	1		1990		
	07 Labor	225	MH	15.50	3 488	
	08 Miscellaneous		LS		122	
						12 400
2	SUBSTATIONS					
	01 Transformer - 5KVA	1	EA	143		
	02 Transformer - 30KVA	2		663	2 992	
	03 Transformer - 75KVA	1		1523		
	04 Labor	78	MH	15.50	1 209	
	05 Miscellaneous		LS		99	
						4 300
3	POWER AND LIGHTING DISTRIBUTION					
	01 1¼" Conduit	130	LF	1.09		
	02 2" Conduit	480		1.75		
	03 2½" Conduit	20		2.77		
	04 3" Conduit	320		6.54		
	05 #8 Wire	168		0.09		
	06 #6 Wire	700		0.13	4 757	
	07 #4 Wire	380		0.19		
	08 #2 Wire	130		0.30		
	09 2/0 Wire	1920		0.68		
	10 4/0 Wire	40		1.03		
	11 350 MCM	80		1.58		
	12 Labor	245	MH	15.50	3 798	
	13 Miscellaneous		LS		145	
						8 700

Fig. 1-7.

Cost-Consultant Estimate

The cost consultant or professional estimator probably has the largest and broadest-based price experience of anyone involved in construction. This is so simply by the nature of his business. He handles the work of many designers and sees a whole range of building types. Usu-

ESTIMATE SHEET

NAME HOW BUILDING
WORK CHANGE ORDER #101

ESTIMATE NO. 1
SHEET NO. 1
OF 1 SHEETS

ESTIMATED BY AB   PRICED BY CD   EXTENDED BY EF   CHECKED BY GH   DATE

QUANTITY	MATERIAL	MATERIAL LIST PRICE	PER	MATERIAL NET COST	LABOR UNIT PR.	MAN HOURS
	ADD Items 1a, 1b, 1c & 1d.					
2'	9" Trench and ends	5 00	E	10 00	2 E	4
22'	12" 2W Trench and ends	10 00	E	2 20 00	2 E	44
23	18" 3W " " " & Horz Ell.			4 00 00	4 E	92
Lot	Engineering and drawings			1 95 00		
	Deduct Items 1a, 1b, 1c & 1d.					
22'	9" 2W and ends	5 00	E	(1 10 00)	2 E	(44)
9'	12" Riser duct	2 00	E	(18 00)	2 E	(18)
1	Wall ell.	17 00	E	(17 00)	5 E	(5)
	Item No. 2					
1000'	¾" EMT Conduit	9 44	C	94 40	12.75 C	1 28
100	¾" Couplings	30 30	C	30 30	4 5 C	45
28	¾" Connectors	17 50	C	4 90	4 5 C	13
14	4" Sq. Boxes & Bushed Cover	60 00	C	8 40	5 5 C	8
	Item No. 3					
0'	3" EMT Conduit	94 51	C	56 71	30 C	18
6	" " Couplings	2 07	E	12 42	1.5 E	9
4	" " Conn & S. I. Bushing	4 20	E	16 80	1.5 E	6
	Item No. 4					
6	6 × 6 × 4 Boxes	1 95	E	11 70	1 E	6
400	¾" Galv. Conduit	22 92	C	91 68	1 5 C	60
12	¾" Locknuts & Bushings	36 00	C	4 32	1.2 E	14
	Item No. 5					
1	39" × 18" × 48" Tel Cab.			1 68 00	12 E	12
8	3" EMT Conn & Bush.	4 20	E	33 60	1.5 E	12
	Totals			12 13 23		4 04

Fig. 1-8.

ally a firm of estimators has some statistical data base upon which they draw in assembling an estimate. They will have available material for the several types of estimates from early or preliminary to final or pre-bid. These people take certain short cuts, combining inseparable or usual costs into fewer groups than may be required by a contractor. They understand where variables lie and attend to these. The example here (Fig. 1-7) shows a page from a professional estimator's final estimate. Labor, shown as a single item, is listed separately on the estimator's work sheet for his own information. In comparing this with the sample contractor's estimate (Fig. 1-8) its simpler approach is evident.

Contractor's Estimate

In addition to its use in formulating a bid price, a contractor's estimate may also serve as a bill of materials for his purchasing department. Frequently contractors' estimates are quite detailed, down to the listing of locknuts and bushings. This might be more evident in change orders such as the sample shown (Fig. 1-8). In this case, of course, the contractor is anxious to build up his side of the case claiming extra work. Change-order work is frequently quite expensive to the contractor because of extra handling of information and materials. It is also difficult for him to charge for this intangible in the form of higher unit costs or quantities, since these are easily verifiable. A complete list of items involved, and sometimes of the special equipment involved, helps to explain the high cost of changed work to an owner or an owner's agent. In a tight bidding situation, however, the result might be different, less detail being required. From the contractor's point of view, the best approach to the bidding table is with as much information as possible on labor and materials for a job. He is then in a position to make intelligent adjustments in his price to suit his posture in the case.

It was previously mentioned that the contractor usually works with the most detailed material available at any time during the design phase. The final or bid estimate is really the only time when an item-by-item takeoff can be made with good accuracy. Labor and materials are priced on a line-by-line basis. The effort serves four purposes:

1. As a bid-price basis;
2. As a materials shopping list;
3. For determination of construction manpower;
4. As a payment application.

Public agencies usually require the early submission by a successful contractor of a line-by-line cost breakdown of the work for purposes of payment (see Fig. 3-14, p. 86). In addition, each line of this breakdown will frequently be required to show labor and materials costs.

Part-Plan Estimate

Another form of early estimating available to the designer or professional estimator is based on the use of a part of each system. After the general outline specifications have been approved and each, or most, of the systems are known, this outline information can serve as a basis for an estimate. A piece of the job that appears to be typical is selected for use. Systems are laid out and detailed in this area, and unit prices per square foot may be obtained from the data. Needless to say, a method of this type should be used sparingly and with an awareness of equipment not appearing within the confines of the chosen space. For an estimate of lighting or wiring devices a fair picture can be obtained. For consideration of feeders this is not the best system.

Chart Method

The charts in this book are intended to allow the designer confidence and choice in placing his design within budget limitations. Each of the money items on the reporting form is obtained from a particular chart or charts showing typical costs for several types. Adjustments in design can be made right at this point, where the effects of more or less luxury are directly visible. There is also provided a parallel set of charts showing the results of choices on the electrical load. Pricing of the work proceeds in the same manner as the design choices. Items grouped as *basic building cost* are the backbone of the electrical design. Price and load selection procedure parallels the route taken by the design engineer. Selection of the electrical loads and entry on the reporting form will provide the user with the size of an electrical service for the designed facility. Service equipment can be priced on the basis of killowatt capacity rather than on a square-foot basis.

The accuracy of any estimate is based primarily on a proper count of each item. Where they are known, counted values should be used in place of those taken from the charts. A situation such as this may obtain later in the design phase, where drawings are relatively complete. At that time the charts can be used as check figures against the estimate of the electrical designer, or tables may be used to establish more specific prices. If necessary, the charts may be used to supplement the information obtained from the quantity takeoffs. This is particularly useful where parts of systems are not complete.

What is called basic building cost herein represents those electrical parts of the building most easily separable as a vital group. This equipment operates together to serve what is electrically essential to permit the building to function. Since this equipment is essentially service, there is relatively little variation from definable standards. Items found in this category are the electrical service, main distribution, feeders, and panels. Also we find that once the lighting level is selected a figure may be placed on branch-circuit work. Wiring devices are a part of branch-circuit work which we find related to building type.

The ability to place this material in cohesive form is not so remarkable as it may first appear. For example, the number of receptacles is directly related to building type. Each building type will have an average room size specific to it (within limits). The number of switches used to control lighting is also related to average room size. The next step is to establish a relationship of these two items to area, depending on building type.

When the number of device outlets has been found, a direct relationship to the amount of branch conduit and wire needed to serve these items can be found. With a relationship determined between the number of devices and the number of circuits required, it follows that branch-circuit panels and then feeders can be priced from the information derived so far. The feeders, once established, and the total electrical load, established from the charts, permit the sizing and subsequent pricing of the main electrical service.

In like manner other parametric relationships have been established. All these linkings have come from the assembly of empirical data, collected from estimates on work designed by a number of design engineers.

Relationships and prices used have been shown to be very consistent in what may be called an average-type building. Of course, immediately one might say the particular project at hand is not an average building, but to a great extent it really is. Generally speaking, where general-construction cost per square foot rises, indicating a higher-quality building, the electrical trade tends to follow in similar ratio. Where there are extraordinary features in a building the designer will be aware of them as part of the program. Such specialties should be carried in the estimate at a premium, and under separate listings.

When dealing with *other systems* there is somewhat less uniformity than is apparent in the basic building. For example, fire-alarm system design is in a great state of flux. The newer smoke-detecting and communicating systems are much more expensive than the older, simpler, manual systems. Even more variation occurs in communications systems such as television. One hospital project had a TV system that looked like it might have been designed for CBS instead of a hospital system. The price (before cutback) was about four times the allowance. What is presented in the charts and tables represents the average cost for a particular building type. The estimator will have to judge where a design has become particularly exotic. Bear in mind that the values reported herein are usable, representative values against which comparison may be made. A building of the type for which the estimate is being prepared has in fact been built for the system price indicated.

ESTIMATE TYPES

There are several types of estimates in use in the construction field. The most dramatic of these and the final test of a set of design documents is the estimate prepared by a contractor to be presented as his bid to do the work. At the complete opposite end of this process we have what might be called pre-design estimates. Along the route between these two ends we may have schematic-phase, preliminary, 50% design-completion, and 100% design-completion or pre-bid estimates. Each of these serves a somewhat different purpose to different interested parties.

Contractor's Bid Estimate

A tendered bid represents an estimate on which a commitment of funds over a period of time is made by the contractor. This is the final cost estimate in the pre-construction process. Of necessity, it should be very close to the amount the bidding contractor will ultimately spend to produce the project. The estimator must deal with the available facts presented by plans and specifications and the imponderables of the markets in the coming years. Of all these data the only hard facts are those shown on the bidding documents. He must make the maximum use of this information. Each item of work or materials has to be examined in detail and tallied as to labor and materials content. We can take off conduit and wire, light fixtures and outlets, switches and receptacles, and the whole gamut of electrical equipment. With ample time spent on this process the physical count will be very accurate. In fact, some contractors do their buying from the estimate sheets. The cost of labor can be fairly accurately estimated from previous man-hour records for particular tasks. Prices of materials are available from suppliers, purchase records, or pricing services.

From this base the estimator moves on to consider the inescapable question of what the future will bring. A projection has to be made through the project duration for the price of materials and for the stability of labor availability and cost. Factors are added, if thought necessary, for the shape of the building, accessibility of the site, local climate, etc. An estimation is made on job overhead costs and applicable home-office costs. Some decision is arrived at on the percentage of profit necessary. The bid package is assembled and complete. All this effort may have cost 0.5% to 1.5% of the price of the electrical work itself (see Fig. 1-9) to produce an estimate of acceptable quality.

Every contractor doing work on other than a negotiated basis is prepared to invest this kind of money in pursuit of business. He also knows that usually the more he spends on an estimate the more accurate it will be. With an accurate takeoff he is free to consider the imponderables of the bidding market in formulating his bid price. A good data base puts the contractor in control of at least some of the factors in a bidding situation. Other factors, such as the number of bidders, the type of bidders, the present state of the market, and the future state of the market, require business decisions, and are normally beyond the estimator's control.

As an aside, consider the effect of this estimate on cost control. This is the most precise and expensive estimate made in the entire design process. In the event that a perfectly comprehensible set of bid documents is produced, there should be very little bid price spread among competent bidders. In this case, the bid price will be the "real" price of the project. At this end of the design process, estimating has very little effect on the cost of the job. In contrast, major cost-effective decisions are made in the earliest stages of design, when

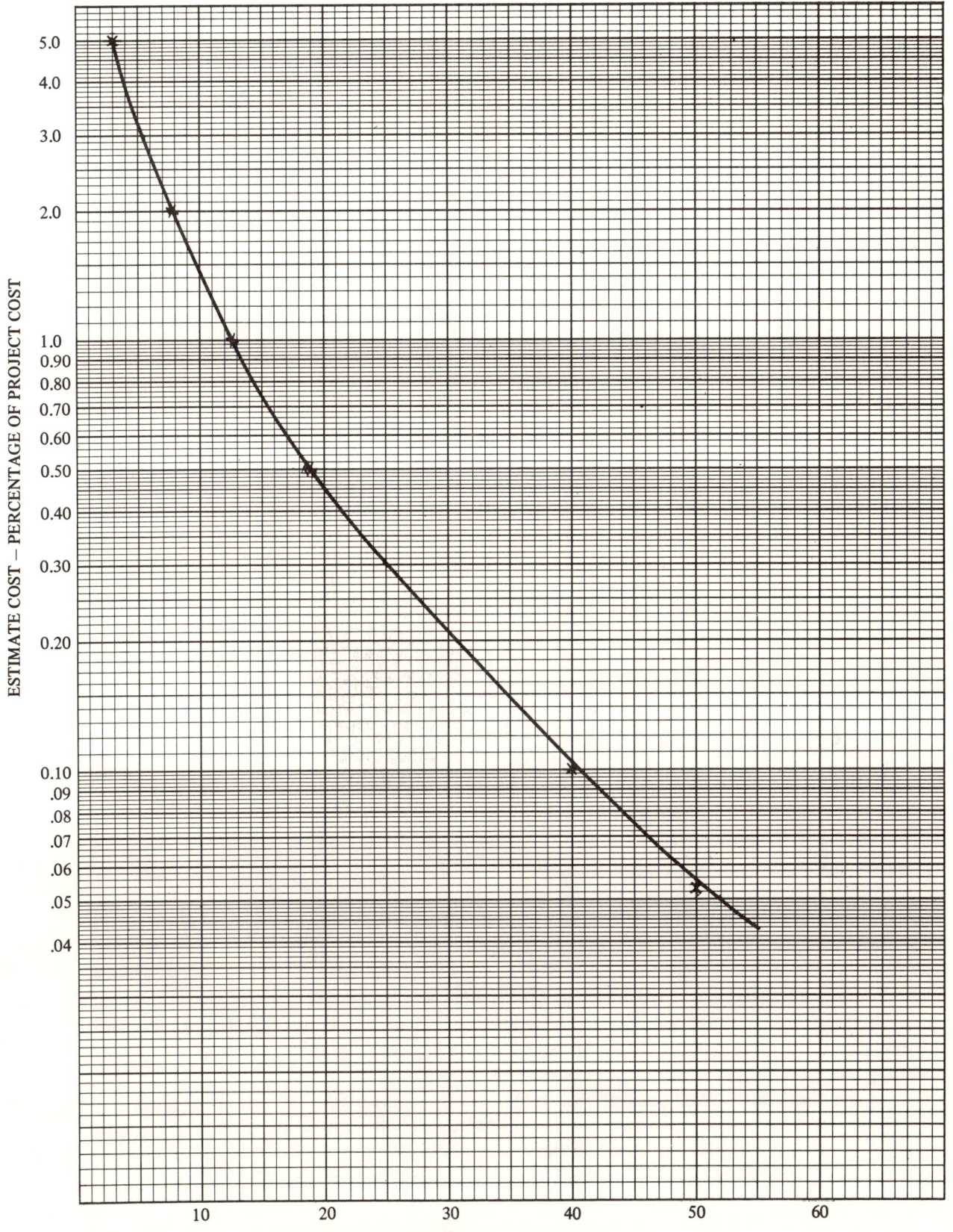

Fig. 1-9.

the least information relevant to costs is available. Cost control through estimating is a process most effective at the beginning of the design work. It follows that early estimates should be given the time and effort they deserve as key guides to the direction final costs will take.

Pre-Design Estimate

At the other end of the design process from bid-document estimating is pre-design estimating. Here, in contrast with the specific information for the contractor's bid estimate, only intangibles are dealt with. There are no plans from which to take off quantities or pricing services on which to base materials prices. We have to zero in by using a series of judgments. An architect or engineer presented with a program must see if available funds can cover the proposed work or adjust the program to suit the funds. From broad indications of scope, some study of economic feasibility has to be made. Schematic design will proceed on the basis of the results of this work, and so on to the more detailed phases of the design process. It is obvious how important these early decisions are. They affect the basic project cost and the extent of the design work.

Where price information is needed most, the least amount of material on which to base cost estimates is available. Things being what they are, the design office cannot spend 1% of the project cost on estimating preliminary design. In fact, the designer cannot spend that amount on his entire estimating expense. Not infrequently this would represent 25% of his fee, an amount that simply cannot be assigned to estimating. Still, for the project to remain economically viable it is absolutely essential that early estimates be accurate. Commitment of funds is based on the least tangible evidence available at any part of the design process. Unfortunately, as so many in the design field know, lack of information is no excuse if the answer is wrong.

Essentially, early-phase estimates are based on historical data. There are several national pricing services which publish data books on building types. These begin to relate costs to several categories: building type, area, cubeage, geographical location, quality, complexity, etc. These price books generally indicate a very wide range in square-foot costs in each category, and so must be used with caution. The best method is to use one's own historical data to compare known results with published price ranges, in order to get a feel for the printed figures. Any surging effects in the labor or materials markets that may influence the project price should be used to adjust that price. Work in each trade should be balanced with the overall design intent and quality set for the building. The use of low-range electrical figures in a high-square-foot-cost general-construction building would not produce proper results. Costs of the entire project cannot be adjusted by changing one or two trades by more than their actual effect on total cost. Price books, with their very wide range of prices, should bear a label saying, "Hazardous—use with Caution."

Schematic, Quarter-Final, 50%, Final, and Pre-Bid Estimates

This group of estimates represents the cost bridge between the pre-design estimate and the proof of the pudding, the contractor's bid estimate. They essentially serve as a cost-control feedback system. Should costs begin to drift up or down during the design process, these estimates can serve to correct that wayward tendency. In the continuum of effect on cost whose ends are at the start and completion of the design, the earlier efforts in this group are the most influential.

Estimates of these types are made on the basis of information available at particular times in the document development process. Where pieces of the work are complete, takeoff can be made and priced on a unit basis. Where less specific information is available, we still have to resort to square-foot prices, outlet prices, per-bed prices for hospitals, and so on. In such cases, prices for systems can sometimes be obtained from manufacturers' representatives, who will on occasion provide price information for their equipment on the basis of discussion regarding performance requirements. They will not usually give installed-cost prices. Costs of installation are taken from the so-called hard data charts and lists of labor and materials once the quantities of the items involved have been established. Bear in mind that specific prices are always better than general prices, and at each stage all the hard data available should be fully utilized.

In spite of the myriad types of electrical equipment required in buildings, price records can be maintained by the design office with an accuracy within that required for estimating purposes. Usually there are general supply houses that will furnish catalogues of a general line of merchandise. Other items can be obtained in catalogues of the nationally known manufacturers. Distributors will have to furnish up-to-date discount schedules with the material lists. As a cautionary word, design offices do not usually have the same safety in "under and over" that contractors' estimators have. Prices tend to rise as the design time passes. Engineer's price data will tend to err only on the low side. The designer must watch that categories of prices are not changing, leaving him with outdated price information. Labor productivity, in contrast to material prices, has remained more or less constant over time. What does change, and in the discrete jumps caused by contract negotiations, is the unit price of labor. This can be handled conveniently so as to adjust costs that have been maintained with separate labor and material components.

ACCURACY REQUIREMENTS

From the point of view of the design office there are two major aspects of an estimate. Primarily an estimate is used in conjunction with the cost model to maintain the project within tolerable cost limits. Estimates also represent an expense to the design office. To put this

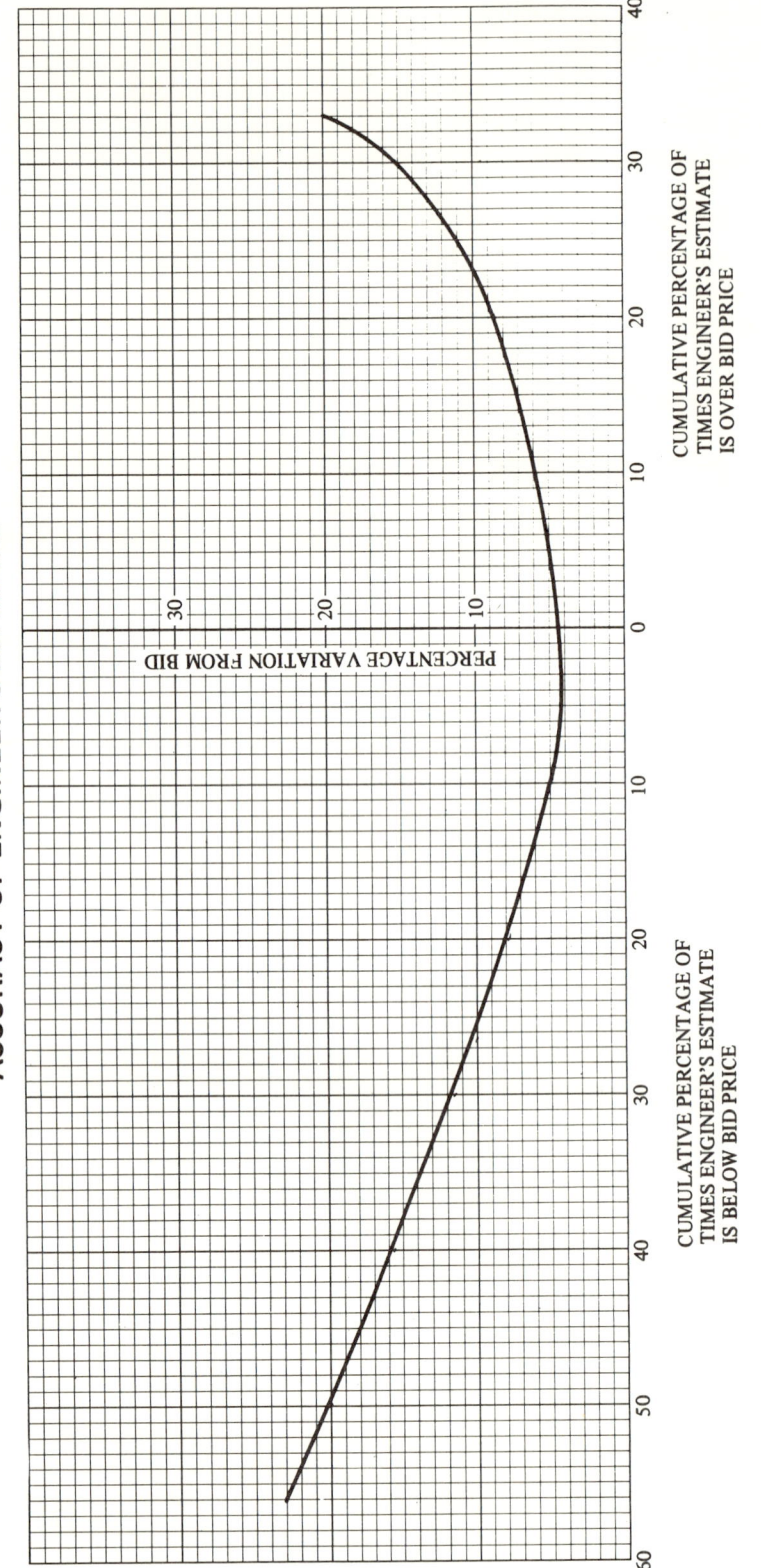

Fig. 1-10.

second aspect on a business basis, the designer should attempt to establish cost effectiveness in his own office for the investment in estimates. In a way, the choice becomes similar to design decisions made in the interest of the owner. A number of design alternatives are considered (as outlined in the section on life-cycle costing in Chapter 2) and evaluated, and a selection is made. Obviously, an infinite number of design choices cannot be considered. So too, in placing a value on the estimating effort, consideration of exposure to risk against the possibility of better financial returns are the factors to be compared.

In Fig. 1-9 the relationship of the cost of an estimate is plotted against its anticipated accuracy. These are the kinds of costs faced by the bidding contractor and not the design office. According to this graph, the acceptable error of 10% is achieved at an expense of 1.5% of project cost. The curve of engineer's estimate accuracy (Fig. 1-10) shows that approximately 45% of the time bid prices come in over the required 10% error limit. The question for the design office is how much to spend on estimates of acceptable accuracy. Here is a quote from *Engineering News Record*, November 20, 1975, page 12, as an example of where uncontrolled prices can go (the article is about the construction of the 1976 Olympic Stadium in Montreal, Canada): "Cost of the 918 × 1,575 ft stadium has now risen to $600 Million, compared to its price tag of $375 Million estimated last July and $100 million as promised by Montreal's mayor, Jean Drapeau in 1972. Lalonde confirmed that the cost of the entire Olympic complex has shot over $1 Billion. The facilities were to cost $126 million originally (*ENR* 6/4/70, P. 15)." Consider the horrendous problems in the design office if a project of this type had a fixed target budget and a contract requiring redesign to budget.

The cost of preparing early estimates in the design office is relatively light simply because there is little information available. More time and its expense will be used in the later phases when drawings become fleshed out. It is then that takeoff estimates are made and require more effort. The reporting forms (Chapter 6) can be used for charting estimates through the whole design process. In this way the same parameters are considered and their waxing and waning noted. The use of the same forms in each reporting period simplifies recognition of problem areas for consultant and client alike. The cost of estimating can be held in line by the use of charts throughout the design process. As the design hardens and areas are completed these areas may be checked in detail and the results may be extended to cover the project. Here the use of several items helps to reduce total error by the principle of compensating errors. The design contingency carried at the outset can be absorbed as specificity, and the confidence it brings, is increased.

Consider the following example of estimating as an investment against loss by the design office.

Example:

Assume the following figures:

Estimated cost, electrical construction = $1,000,000
Design fee (4% of electrical construction cost) = $40,000
Low bid = $1,200,000

Overage $\dfrac{1,200,000 - 1,000,000}{1,000,000} = 20\%$

The usual maximum over budget that any client will accept is 10%. Frequently the number is less, but 10% will be used for purposes of illustration. To cut back to the acceptable 10% over budget or to $1,100,000 requires a reduction of 8.3%. Things being what they are in the real world of rebidding, the designer should plan for at least a 15% cutback to insure the final price. What will this effort cost the designer?

Redesign work is more difficult than work on the initial design. Frequently entire systems need reworking to incorporate small monetary changes. This work then is worth approximately 1.5 times original design work in production costs. In this case:

Design fee = 4%
Design cost = 2%
Redesign cost = 1.5 × 2% = 3%
Required project cost reduction = 15% × 1,200,000 = $180,000
Redesign cost = 3% × $180,000 = $5,400

Thus, $5,400 is the expense required on the part of the electrical designer to redesign the project for the required reduction in project cost; as a part of the total design fee, this amounts to

$$\dfrac{5,400}{40,000} = 13.5\%$$

In other words, the penalty for poor estimating or poor cost control in this case is 13.5% of the entire design fee. (This is not to mention the intangible loss in project delay and client grief.)

On a project basis the redesign cost is equal to:

$$\dfrac{5,400}{1,000,000} = 0.54\% \text{ of project cost}$$

Figure 1-10 shows that a contractor can buy an estimate of approximately 18% accuracy for a cost of 0.54% of the project cost. Since in the illustrative example above it took 0.54% of project cost to correct a 20% overestimate we have confirmation of the order of magnitude of the expense involved.

This establishes an approximate upper limit on the cost of redesign against the cost of a good estimate. Since the designer does have control over costs and the estimates are less detailed, they do cost less than those prepared by contractors. On this basis a figure of up to 10% of the design fee may be considered the offsetting investment in good estimating against potential loss. These expenses may be made either in the office or to an outside cost consultant. In the example used for illustration, a $2000 to $4000 expense for estimates through the design process can be seen to offset a possible expense of $5400 for redesign.

If the designer goes outside his own office he can find cost consultants available for fees similar to those which his own office expense might be. This may be expected to come to a figure around the 5% mark in terms of the design fee. For this number a designer can buy a cost-control package from budget phase through the bidding process. A good estimating and cost consultant will help in pricing available choices to design solutions. Used in close cooperation with the design office a cost consultant can be a useful adjunct to the staff. There is also the intangible advantage of relieving the designer of some of the worry over cost control. However, if the cost-control services are not physically convenient to hand and are not used early and often, much of the potential advantage is lost.

The choice of a cost consultant should be made with care. A visit to the consultant's shop and an interview with key personnel is essential. Consultants should be more than takeoff men; they should be people with professional training and experience. Since part of the designer's staff will be in contact with the consultant's staff during the design process, a certain amount of sympathetic feeling should exist between the two groups. A good consultant should be able to guide the designer along the path the designer wants to follow, but will serve to remove the cost pitfalls. If the design staff is unable to work in this manner with outsiders, forget it, but don't forget that cost control remains an item of expense for the designer's office.

2. Cost Control

SUMMARY

Cost control is an activity that should start in the design office and continue throughout the life of the project. An important method of controlling costs in the design phase is by means of a cost model and its periodic updates. These updates are based on phased estimates through the design phases and on a consistent method of presenting the information to the owner.

Value engineering is another cost-control system. It is a reviewing technique usually employed by people not a part of the design team. It too is based on sound estimating requirements.

In order to view the total cost of a construction project over the life of the project, a technique called life-cycle analysis is employed. Not surprisingly, it too requires good estimates of alternative solutions to the design problem.

Cost control: These have become two key words in the language of construction, with the increasing building complexity, the rise of multidisciplinary technology, changing labor rates, escalation at one rate of one product and deescalation at a different rate of another, and so on. At first introduction, the problem of producing construction for a particular price is mind-boggling, and on sober consideration it remains mind-boggling. The situation is not, however, completely hopeless. Systems of cost control have been devised and do work. Costs can be maintained from design budget to construction.

Public and private clients are becoming very cost conscious. The day of the open-ended construction cost has passed. Responsibility for design to cost has been placed squarely on the backs of the designers, where it rightly belongs. Typical design service contracts with the U.S. Postal Service do not mention an allowed variation of bids from design estimates. The New York State Dormitory Authority typically demands construction bid prices to be within 5% of design cost estimates. Construction for New York City normally will not proceed when prices are 10% or more above design estimates. This pricing concept is also recognized in the AIA (American Institute of Architects) contract, 1974 edition. Under article 3, "Construction Cost," paragraph 3.5 states: "When a fixed limit of Construction Cost is established as a condition of this Agreement, it shall be in writing signed by the parties and shall include a bidding contingency of 10 % unless another amount is agreed upon in writing." Later, in paragraph 3.5.2 it states: "the Architect, without additional charge, shall modify the Drawings and Specifications as necessary to bring the Construction Cost within the fixed limit." Since more often than not the design engineer working on the electrical phase of the project is bound by the same conditions as is the Architect; he will also be required to design to cost.

Note that control can only be maintained by frequent monitoring. Costs are most sensitive to control early in the design process, when the major decisions are being made. Later on, when the design is locked in, changes become minimal and so do their effect on cost. The larger concepts determine how the bulk of the money is spent. Frequent, early, realistic cost comparisons are an absolute necessity for keeping a project on its cost track. The early effect on costs is shown graphically in Fig. 2-1, borrowed from the Value Engineers.

By the time a project is being detailed by the draftsmen, little effect of a noncatastrophic nature can be made. A corollary to the timeliness of cost-effective design choice can be seen in the effect of a change on the cost of the design work itself (Fig. 2-2).

Presumably the design office is in business to make a profit. Many contracts for design services require the final cost to be within budget or a redesign must be effected at no additional fee. The time to make any changes least financially damaging to the design office is early on. When drawings are covered with lines, fully detailed, and cross-referenced, it is no time to make changes. To use the graphical analogy again, it can be seen that the cost of changing drawings is lower the earlier in the process these changes are made. Happily, in this case the result is that the most effective time to maintain a client's budget is also the time when this effort costs least.

If we expand the design portion of the value engineer's curve and put it on the same coordinates as the designer's production cost

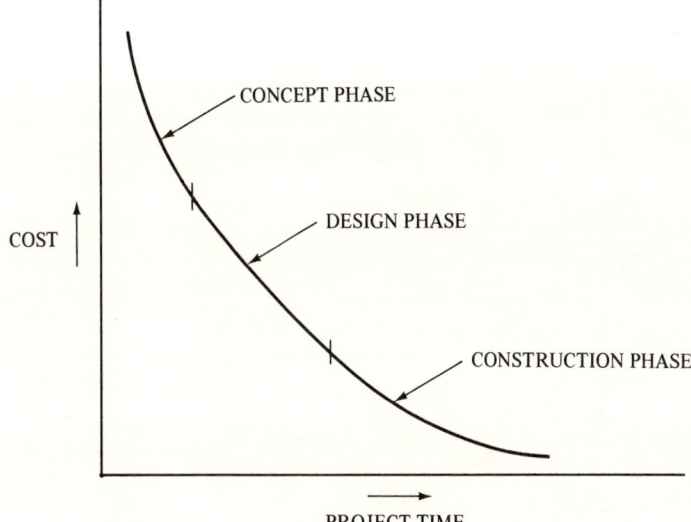

Fig. 2-1.

DESIGN CHANGE COST

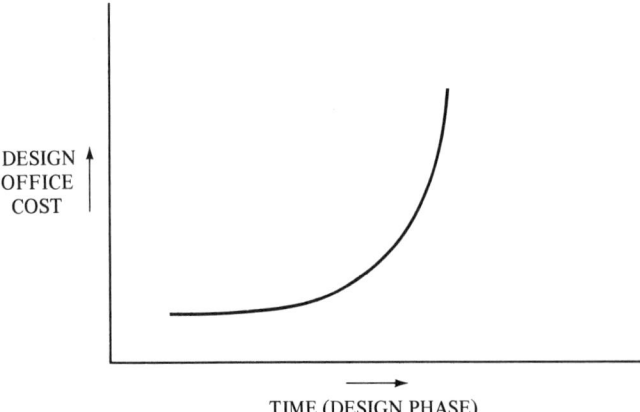

Fig. 2-2.

curve, we get the type of picture shown in Fig. 2-3, demonstrating the large return for small expense early in design and the reverse later in the design phase.

It is evident that in actuality we are looking at cost control from two aspects: first, is the construction budget requirement; and, second, the requirement to operate the design office on a profitable basis. Normal practice in the industry indicates that the best, that is, the most accurate estimate of project cost is made from the most detailed set of construction documents. These estimates also are better if larger amounts of time are spent on them; put another way, the more expensive estimate is more accurate. Again we can resort to simple illustrations to demonstrate these points.

The curves in Fig. 2-4 appear to demonstrate a dilemma for the design office. What is being indicated is that great expense must be spent in using finished documents to produce accurate estimates. A process such as this would be intolerable for a design firm. The de-

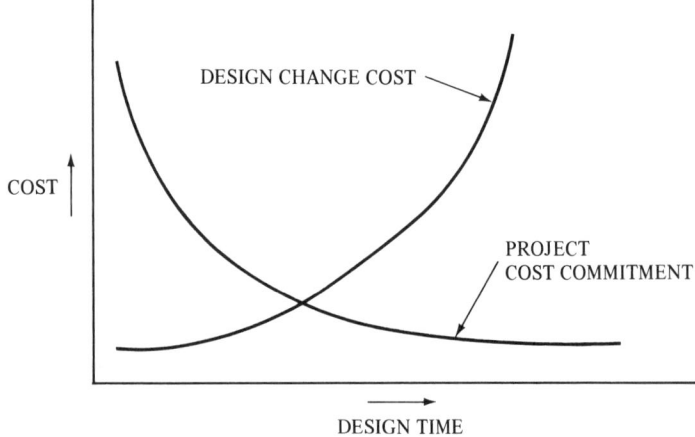

Fig. 2-3.

ESTIMATE ACCURACY VS DOCUMENT COMPLETION

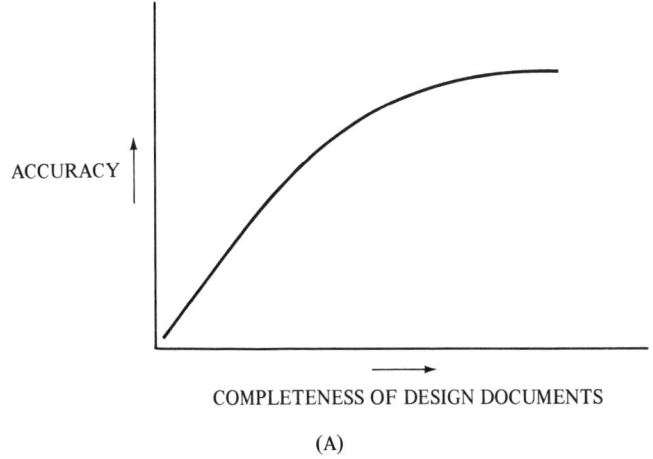

(A)

ESTIMATE ACCURACY VS COST OF ESTIMATE

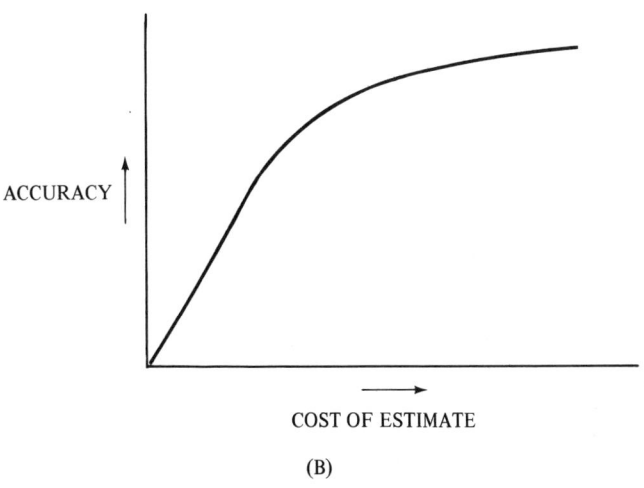

(B)

Fig. 2-4.

sign firm cannot wait until documents are far along toward completion to make estimates and the redesign that excessive indicated costs would then require. Nor can the firm spend large sums on estimates unless this is made part of their fee structure. It is true here as in many aspects of the construction process that some phase must be viewed from more than one point of reference. It is often essential to produce a design to meet a specific budget while this effort in itself is a cost to the design office.

Awareness is the watchword in solving these problems. Very rarely does project cost decrease during the design process. Greater detail somehow seems to bring greater cost. The program budget, schematic-phase budget, or whatever one calls the earliest cost statement, should be evolved with great care. Really critical examination should be made of the necessity and cost of each line item. It is at this time that the large blocks of economic choice are put in place. Costs are assembled, and a design contingency is carried against later imponderables rounding out the estimate at this phase. Subsequent estimates can carry a reduced contingency as the construction drawings and specifications become more complete.

Construction projects vary in complexity. The difficulty of accurate estimation and the cost control it requires will be affected by the complexity of the project. It can be readily appreciated that a simple warehouse can be more accurately and more easily estimated than a school or a hospital. The simpler requirement for estimates of simple buildings occurs at each phase of the design process. Using the usual square-foot check figures will produce a more accurate estimate for the warehouse than for the school at any stage of design. To put this another way: the final test of estimates are bid prices, and these will vary more widely for a complex building than a simple building. A designer must therefore pay closer attention to costs in a more complex project.

Bid-price range in a simple building type may be ±10%; a moderately complicated building may see this range widen to ±15%; while the most complex types may produce a range of ±25%. With this type of variation, cost controls on the more complex projects are most important and most effective.

At this time we should examine cost control from a quantitative standpoint. In this view, the successful design produces bid prices which enable the job to be bought at the budget price. Within narrow limits, usually ±10%, bid prices that are too high or low may permit a project to be built, but higher prices necessitate redesign. On the opposite end of the scale, lower prices do not take proper advantage of available funds. In other words, the design cost has to be within the normal bid spread for a project to get built without design change.

Recognizing that simple work can be estimated more accurately than complex work, we have broken our estimate format into several categories. In this manner simpler items are separated out of the whole project to take advantage of the possibility of more accurate forecasting of costs on the simpler work. As a result, we have reduced the cost spread on a substantial part of the project's electrical work. If we were to price the entire building on a square-foot basis, our accuracy would be considerably less than that produced by a sum of the more accurately estimated parts, as shown in the example below.

Example:

Hospital Estimate
Cost per square foot basis:

    Accuracy 25% over the entire project;

    Total project accuracy: 25%

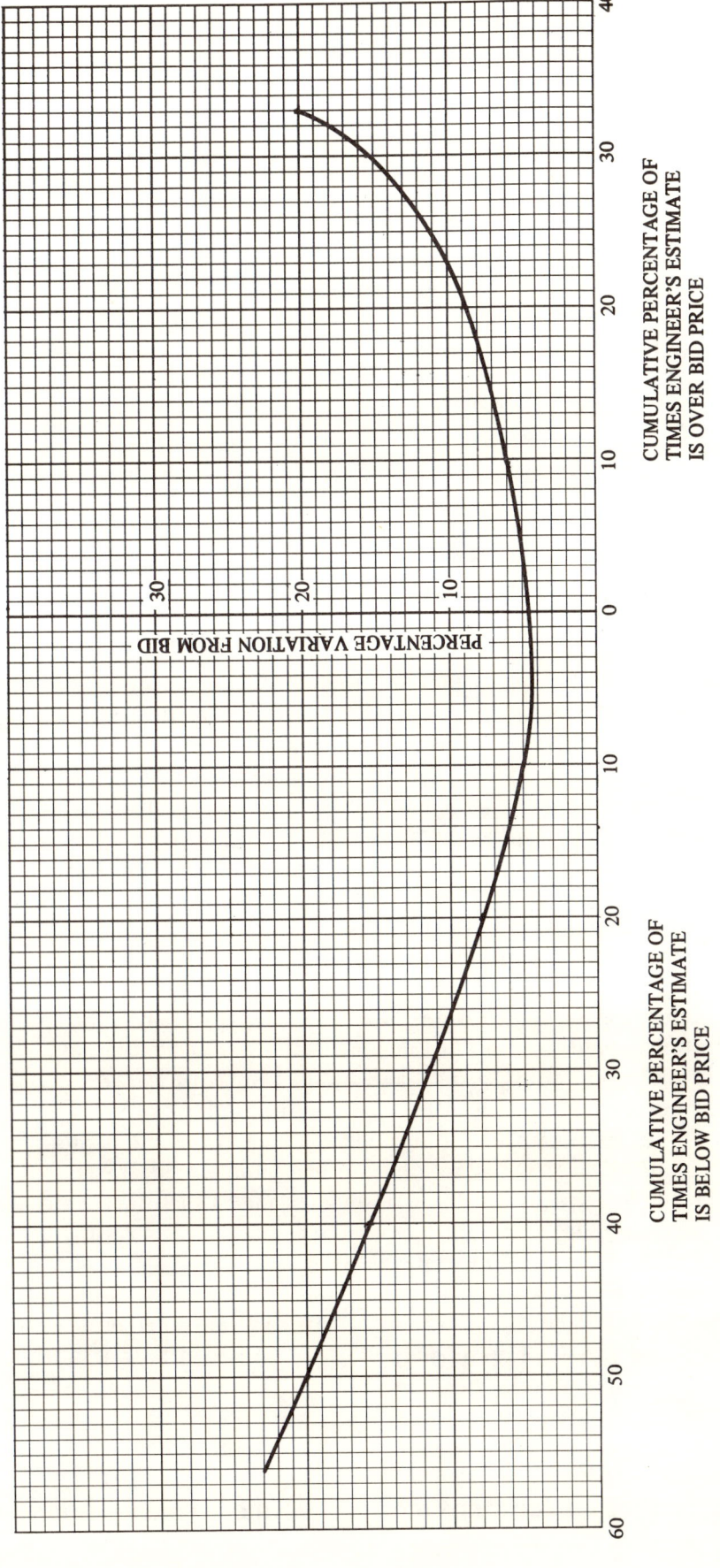

Fig. 2-5.

Sum-of-parts basis:

Part A: 25% of project, accuracy 10%
Part B: 35% of project, accuracy 10%
Part C: 10% of project, accuracy 5%
Part D: 30% of project, accuracy 25%

Total project accuracy: 14%

A number of actual bid prices collected over approximately a year are plotted against estimated values for the same projects in Fig. 2-5. Building types were mixed and projects ranged from quite small in size to several million dollars. These were put on the graph to show a distribution of under- and overestimation. We find almost 54% of the projects at a deviation of up to 10% over or 10% under the estimated price. Almost all the projects were within 20% of the estimate. Of course, there were excursions. Contracts written with public agencies may require redesign if prices are 10% or more over budget, and some even talk of a 5% tolerance. From our illustration it appears that about 38 of the time bids were above the acceptable 10% override. From the point of view of the designer's economics this must be prevented. We also see from the shape of the curve that more estimates are below bid price than above, confirming our opinion that costs tend to rise as design drawings are completed. What we are saying here is that only about half the architect/engineers are putting out designs that have been properly designed and controlled. We assume that the others will be in poorer financial condition.

As a demonstration that cost control works, study Fig. 2-6. Plotted on the same sets of axes are the bid results for four Post Office buildings. The four buildings have similar requirements but were designed by four separate design offices. The one project with good bid spread was done by the U.S. Army Corps of Engineers, who exercised strict cost control during the design phase, and produced a complete set of comprehensible documents. More than 60% of the bidders were within 10% of the low bid, whereas the other three sets of results were quite poor by this measure. Breaks in the curve may indicate different interpretations of the documents. Here again the least sharp break belongs to the plans drawn by the Corps of Engineers.

With poor documents and/or poor cost control the risk of redesign expense by the design office is high. If you look at the two poorest results in Fig. 2-6, you can see that on one there was a 10% jump between the low bidder and the next bidder, and in the other there was a 20% jump. If the low bid is not close to the budget mark there is a high risk that the 10% permissible variation will be lost in such jumps.

Figure 2-7 represents bid results on a single cost-controlled project. Electrical and HVAC bid results are plotted. Although both results were within the budget price, the electrical estimate being 3.5% above the low bid and the HVAC estimate being 5.4% above the low bid, the results show superior documents and/or control for the

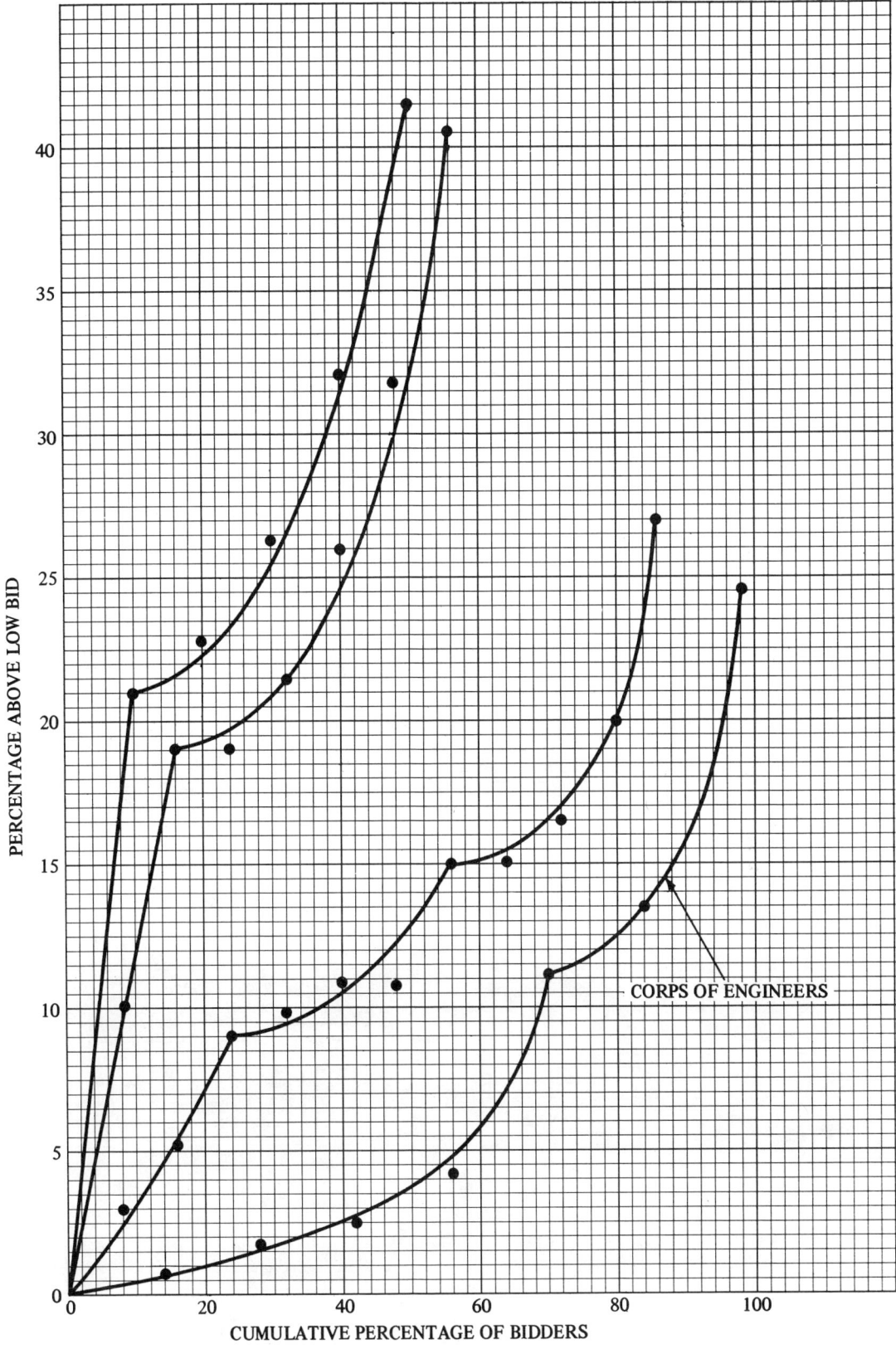

Fig. 2-6.

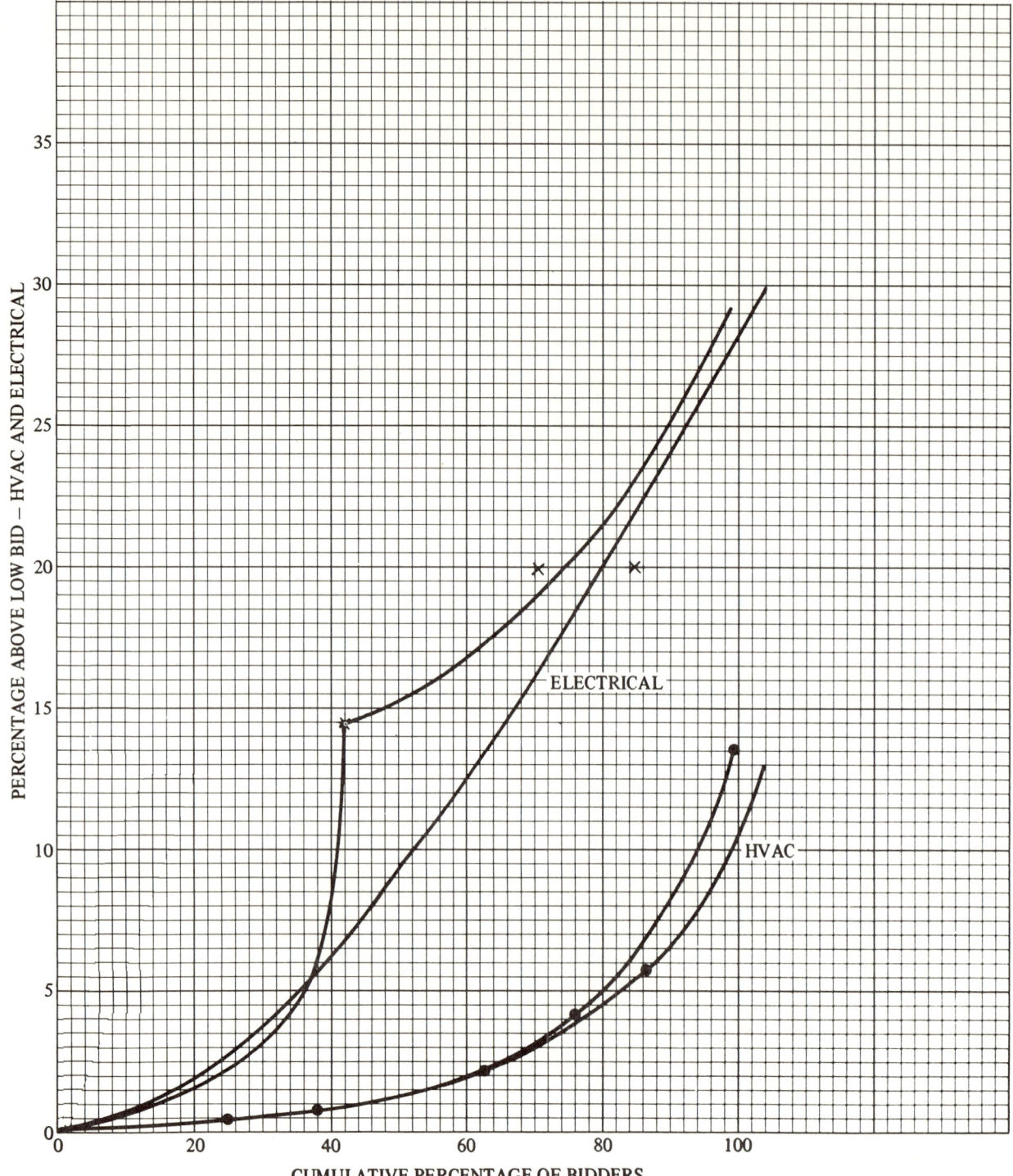

Fig. 2-7.

HVAC drawings as compared with the electrical drawings. Both of these trades were bid with a comfortable number of bidders below the 5% variation allowed. In the case of the electrical work, approximately one-third of the bidders were below the line. HVAC work bids showed about 80% were below the 5% requirement.

In addition to bidding-document completeness there are a few other factors affecting the bid price received. The designer has somewhat less effectiveness in controlling these factors:

1. Cost of money
2. Bidding climate
3. Job duration
4. Job management.

As interest rates fluctuate, so will the price of a job. There is relatively little, aside from selecting an earlier or later bid date, that the designer can do to control this item.

Bidding climate to a great extent depends on local market conditions. Good times produce fewer bidders and higher prices, while bad economic conditions produce more bidders and lower prices. The designer should attempt to interest as wide a group of bidders as possible in bidding his job.

A long construction period places the strain of certain imponderables on the long-range projection of material and labor costs. Economic forecasting has become very tricky recently. In addition, long construction periods require additional job and office supervision, which represent real costs.

Job management is very important, but can not adequately be evaluated at the time of the bid. An efficient construction manager can really make a difference in the contracting costs by proper scheduling and good job administration. If the CM is chosen prior to the bid, his reputation may have some influence on the price.

Cost Model

Let us assume that the overall budget has been determined for the project and a part has been assigned to electrical work. This single cost item can be broken down into smaller parts for more accurate control and cost isolation. An assemblage of this information can be called an *electrical cost model* (Fig. 2-8). The model once assembled should be periodically updated for cost control. As the project moves from the program stage to the detailers, changes and shifts in cost can be observed before too much damage is done in the form of time commitments to designs which cannot be contained within the budget. The cost model here is set up to utilize the estimating system described in later chapters. However, although costs will change, the category designations remain the same to the end of the design process and line items in the cost model are adjusted to suit. As the job progresses out of the designers' hands and into the detailers' province, estimates based on quantity takeoff (as the more specific determinants of cost) should be used to update the cost model.

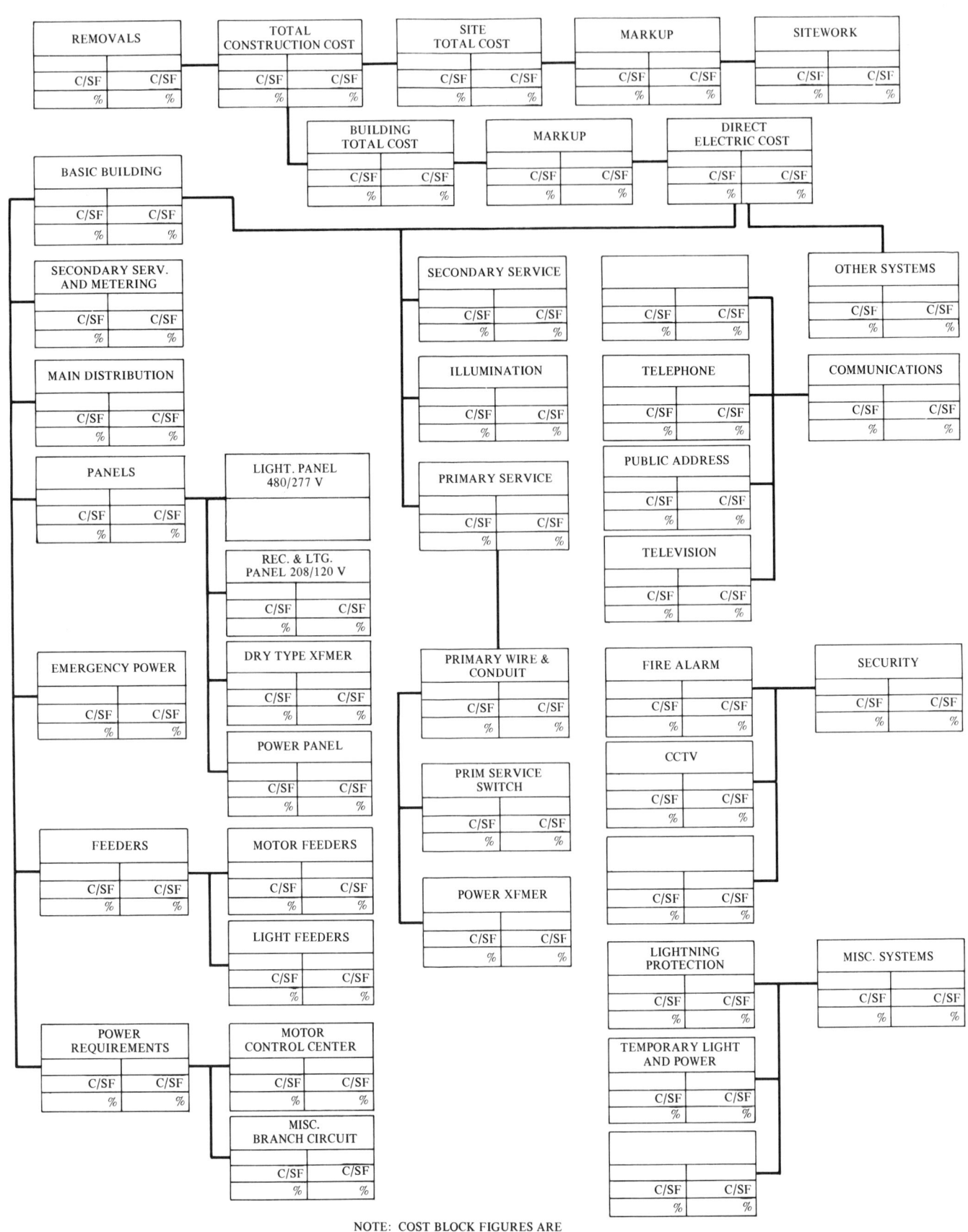

NOTE: COST BLOCK FIGURES ARE WITHOUT OVERHEAD AND PROFIT

Fig. 2-8.

The cost model is essentially a shorthand method for assembling cost data into a convenient single picture. Each of the cost blocks used here should contain two figures. One amount should be a possible amount and percentage of the total electrical cost, while the other should represent the estimated cost of that block and percentage of the electrical cost at the moment of the estimate—in other words, the actual amount. These figures are all taken on a dollar cost per square foot (C/SF) basis without markups, to be compatible with each other and with the cost figures given elsewhere in this book. For example: the lighting block would look like this:

	Lighting	
POSSIBLE C/SF %	P	A
	0.85	1.35
	30%	40%

POSSIBLE C/SF %   ACTUAL C/SF %

The sum of the possible blocks will be set up to represent an acceptable total cost and cost distribution for the project. Figures to be used for possible blocks can be taken from the charts showing typical cost breakdowns by building types. Totaling the actual blocks will give the total electrical cost. Actual figures, of course, are those taken from the appropriate places in the estimate form given later. Cost deviation from permissible levels can be quickly spotted by this method, and corrective action can be taken if needed.

Blocks set up to show the percentage of the block contents in the overall electrical work serve two purposes. The first, as we have seen, is to make a reasonable and workable assignment of costs to the components of the electrical work. The second feature permits comparison with other projects. Within the same building types, costs can vary rather widely. Frequently, a price ratio of 1:5 can be found in a single building type. However, the importance of each of the cost blocks in relation to the other cost blocks remains fairly typical within a single building type. Where some special feature is made in the projects, its cost block percentage should rise and appear exaggerated. If this occurs in an item not intended to be of special importance, a warning has been sounded. The designer may then investigate, validate if the change is justified, or adjust his design if required, to bring the block back into proper dimension.

Charts used for estimating will permit an exercise of choice in selecting items for each of the cost blocks. Should total costs at any one time exceed allowable costs, the charts may be consulted for possible cures. If early estimates indicate a problem which cannot be corrected by adjustment of the cost blocks, a reexamination of the program will have to be made. This problem will usually yield to approaches made by Value Engineering techniques (see discussion of Value Engineering later in this chapter). Left uncontrolled, costs have a tendency to rise.

The cost model in this book is based on raw costs for the items it contains. These are costs with no markups, no escalation, and no contingencies or factors; that is, these are the costs as taken from the cost charts in Chapter 5. The reason for using these figures lies in the basic intention to be served by the cost model. This is to discover locations of excessive cost. Since the factors applied to each line item of cost to obtain the final estimated figures are all the same, they will not alter cost figures in the cost model boxes in a significant way. The cost model serves to compare a project under consideration with historical data. To do this, both projects must be on a comparable base.

In the process of estimating a project, the estimator will obtain prices based on today's market. He becomes familiar with these prices and recognizes exceptions and errors when they occur. This is an additional reason for basing the cost model on raw cost data.

Of prime importance in this or any other comparison operation is a clear statement of all conditions and assumptions used in arriving at the cost figures. This may be obvious, but more often than not one sees estimates that do not make the factors clear. When you consider that factors such as general conditions, overhead, profit, design contingency, construction contingency, and escalation may well amount to 50% of the raw costs, the importance of a clear statement regarding the use of these factors becomes clear.

The practice of using raw costs in the line items of the estimate and then marking up the total is also a labor-saving device in processing the estimate. Assume, for example, that a project has been delayed in starting or that the construction period has been extended; to revise the estimate, one merely has to adjust the escalation multiplier, the line items remaining the same. Note that a change of this nature will require very little adjustment to the cost model. The individual cost boxes will not be affected at all; only the adjustment and total-cost boxes will change.

This cost model has an additional feature; a place to enter the percentage of the cost subtotal contained in each cost box. This can serve to highlight cost dislocations where an entire project may be exceeding the model budget. In such a case, disproportionate costs will show up as unusual percentages. Our cost model serves the purpose of pointing to areas where Value Engineering techniques may be applied by viewing percentage indication and the raw-cost comparison.

The main use of the cost model is as a tool for highlighting locations of excessive cost. There are reference units other than the cost per square foot used in this book that may be used to accomplish this. For example, the EPA (Environmental Protection Agency) uses a base of gallons per day in their cost model for sewage-treatment plants (Fig. 2-9). In this case, the total electrical cost is represented by a single cost block. The illustration shows an 88% cost overrun in the electrical work, but will need additional breakdown to isolate the cause of the overrun. Cost models could be constructed on cost-per-

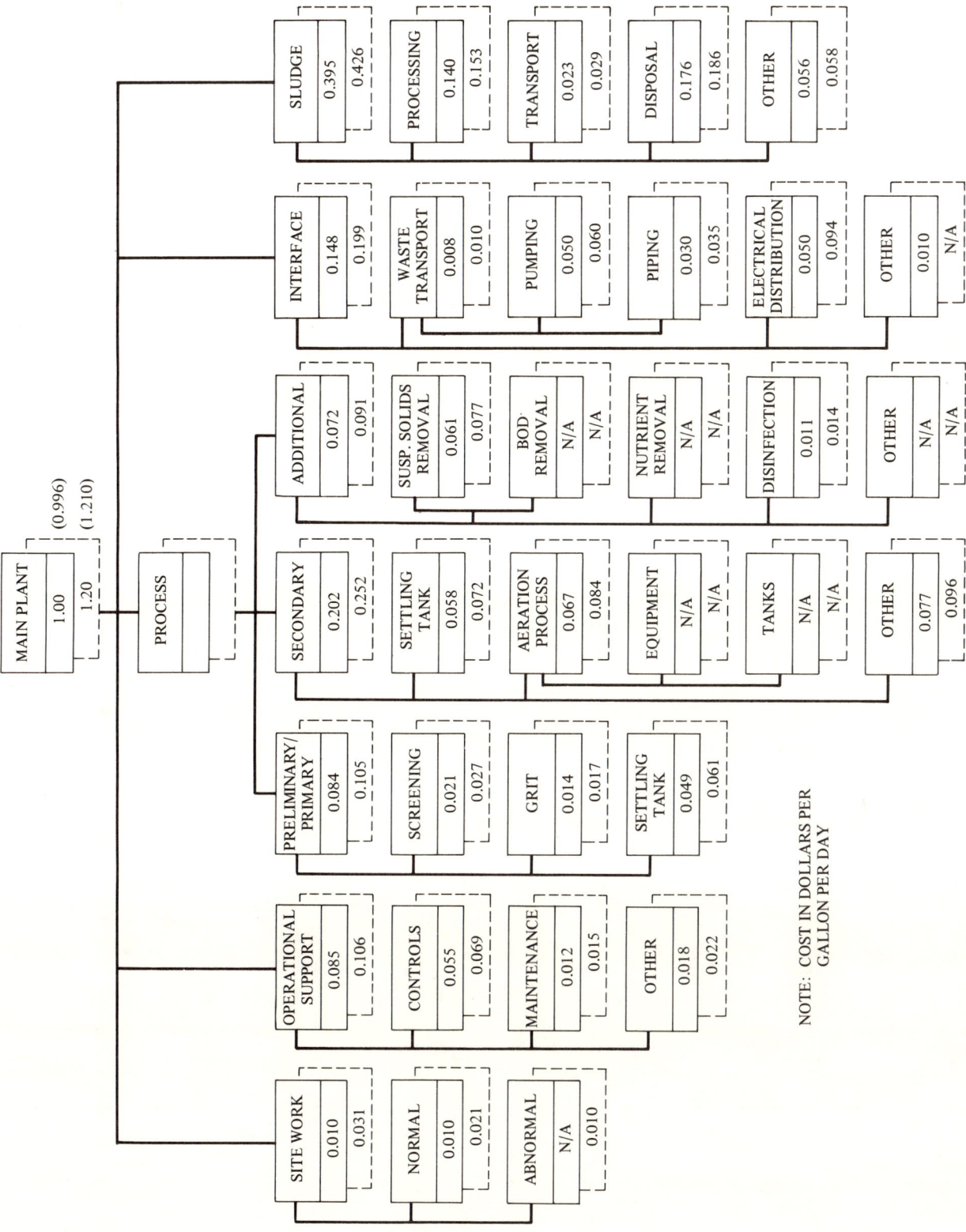

Fig. 2-9.

COST CONTROL 45

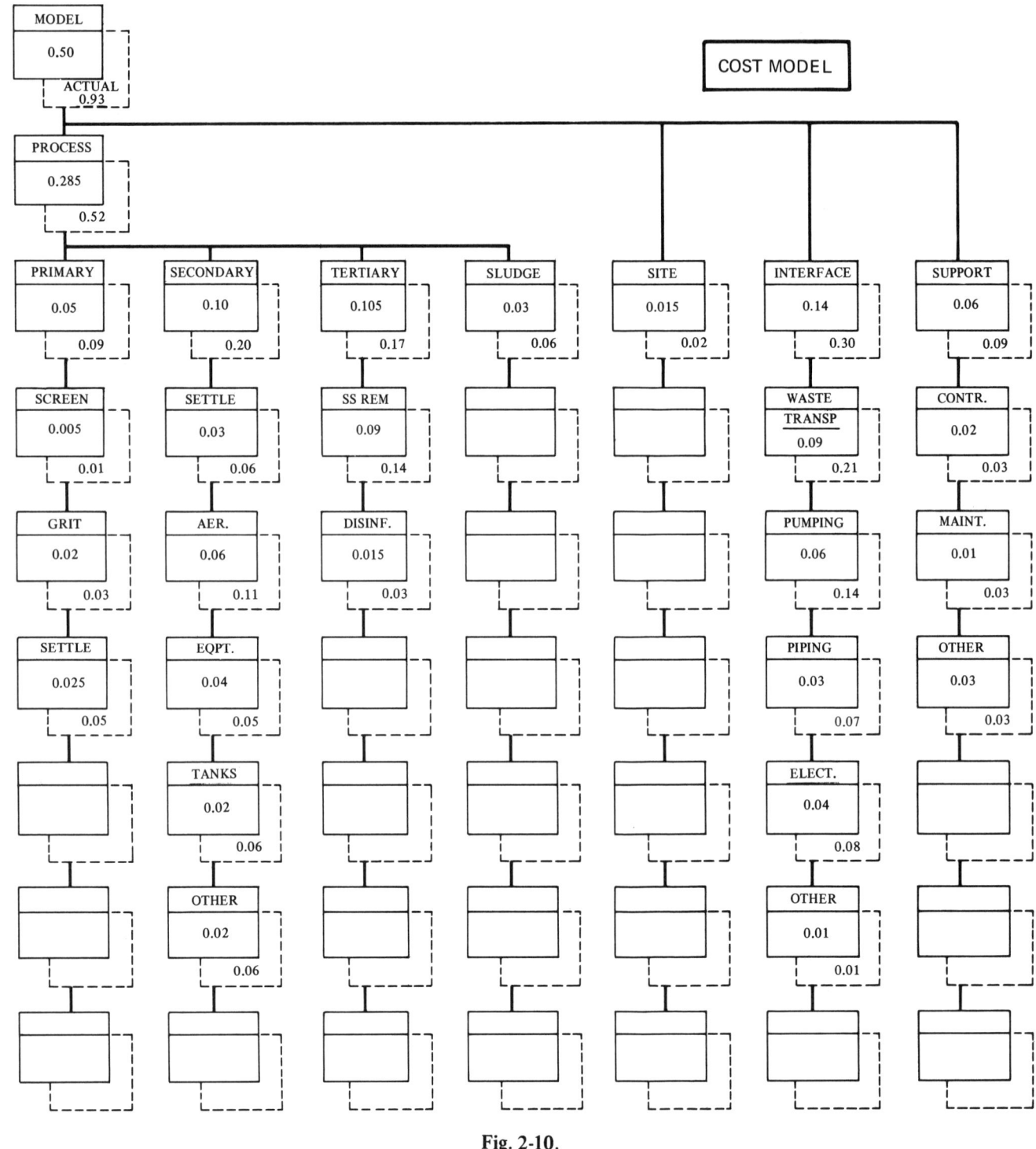

Fig. 2-10.

ELECTRICAL COST MODEL

		Possible		Program Actual		Schematic Actual		Prelim Actual		50% Actual		Final Actual	
		CSF	%	CSF	%	CSF	%	CSF	%	CSF	%	CSF	%
1.0	Service & Distribution												
1a1	Primary Wire & Conduit												
1a2	Primary Switch & Fuse												
1a3	Power Transformer												
1b1	Secondary Service Wire & Conduit												
	Basic Building												
1b2	Secondary Service & Metering												
1b3	Main Distribution Equipment												
1c1	Lighting Panel 480/277 V												
1c2	Light & Receptacle Panel 208/120 V												
1c3	Dry Type Transformer												
1c4	Power Panel												
1d1	Motor Feeders												
1d2	Lighting Feeders												
1e	Emergency Power												
2.0	Power Requirements												
2a1	Lighting Outlet Box												
2a2	Receptable												
2a3	Light Switch												
2a4	Branch Circuits												
2b1	Small Motor & Control												
2b2	Large Motor & Control												
2c	Motor Control Center												
2d	Kitchen												
3.0	Illumination												
4.0	Other Systems												
4a	Communications												
4a1	Telephone												
4a2	Public Address												
4a3	Television Antenna System												
4a4	Clock System												
4b	Security												
4b1	Fire Alarm												
4b2	CCTV												
4c	Miscellaneous Systems												
4c1	Lightning Protection												
4c2	Temporary Light and Power												

Fig. 2-11.

kilowatt, cost-per-bed, cost-per-student, cost-per-meal, etc., bases, depending on what the electrical designer feels will be most significant for the project under consideration. Once the Value Engineering or cost correction has been performed, cost items may be converted back to the common base used for the project as a whole.

Another EPA cost model (Fig. 2-10) illustrates general construction costs for a part of a treatment plant. In this, electrical work is considered part of a category called *interface* in value engineering terminology. The electrical block will require further detailing of this block, as previously discussed.

Another form the Cost Model may take is the somewhat less graphical, but perhaps all the more convenient, shape of Fig. 2-11. Here the single form tabulates the results of the periodic estimates for a ready item-by-item comparison. This type of form may be issued at the intervals required by contract or in the event of an emergency cost dislocation.

Periodic Estimates

As previously discussed, there are critical times during the design phase when estimates should be prepared and received. These are presented for client acceptance in the same form each time to permit line-by-line comparisons. The most significant changes can be readily identified and then dealt with.

Estimates used in this manner are:

1. Program
2. Budget or schematic phase
3. Preliminary or early design
4. 50% document completion
5. 100% document completion, final, or pre-bid

All these estimates should contain the required elements of project cost. These are fully discussed elsewhere:

1. Design contingency
2. Escalation
3. Market conditions
4. Job-difficulty factor
5. Contractors's overhead and profit

1. Program Estimate

This is the first application of money as a yardstick of reality to measure the scope of a proposed construction project. Under consideration at this time will be:

1. Type of building proposed;
2. Quality of construction—institutional, federal, state, speculative, etc.

3. Site location—for relative difficulty of site preparation, worker access, storage facilities, material delivery conditions;
4. Nature of the local labor market;
5. Timing of construction period relative to local construction activity.

Prices used for this type of estimate are usually based on square-foot costs. The best guide is a similar project designed in one's own office. Then there might be several not so similar but still relevant projects, forming part of the office historical file. Prices may be scaled up or down to suit the new project size or current economic indicators. The cost-capacity method can be used where an office has developed a good set of cost-capacity factors. Construction-cost services by several publishers provide periodic price sheets or annual price books which can give a guide to cost range. Any of these methods will give a rough cost per square foot.

Charts presented in this book will also serve to provide early rough costs. In any event, whatever, the early method used, the reporting forms included here assist in breaking down even the earliest cost considerations for a more precise and reliable program estimate. Figures produced from this effort should be discussed with estimators for the other trades and with the project manager. Program feasibility begins to develop from these early costs. Items that need trimming, or will allow expansion, become apparent, as does the outline of cost-control requirements. A cost model for the electrical work and other trades should be prepared from the data collected. Project costs are assembled from these square-foot costs by adding the various project-determined cost items.

2. Budget Estimate—Schematic Phase

Having an acceptable cost-to-program relationship established permits the development of schematic drawings. Architectural drawings at this phase will have the building configuration, space assignments, sections, some room layouts, and some general data on site requirements. All the design trades should have produced generalized one-line diagrams and systems-outline specifications. Requirements for air conditioning, kitchens, and specialized motor loads should begin to appear. Electrical heating requirements can be developed. Site lighting requirements will turn up. The scope of a lighting system should be evolved.

Where information is now quite specific, the estimating techniques of takeoff and unit pricing may be applied. Small areas can be developed according to requirements, and unit prices developed for them may be applied across larger areas. Other prices may be taken from the charts. The charts will permit ready identification of layout adjustments (for example, of foot-candles or outlets per square foot) where costs demand them. Downstream changes will of course affect items all the way back to the service. The charts should readily provide figures showing the full range of effects due to branch load changes. These figures are again assembled in the reporting form and the cost model. Project costs are added as they apply, and figures are reviewed with the client before proceeding. It might be noted that the cost efforts of all trades should be considered in order to see which way the total effort is going. Changes by the client in scope or available funds may occur at any stage of the design process. These should be dealt with as soon as possible to avoid later misunderstanding.

3. *Preliminary Estimate—Early Design*

Having passed through the more formative, early stages, drawings of all trades are directed toward their final form. Some of the equipment has been selected and located. Architectural layouts, sections, and fenestration are effectively completed. Room layouts are being developed. Special areas are being worked. Space assignments for equipment pipe shafts and horizontal mechanical zones are defined. Some of the motorized equipment begins to appear in the schedules, and some of this is located in plan. Lighting design is set, at least at the foot-candle design level. Light and power panels can be located and sized approximately. Other electrical systems such as security and communication can be generalized. Site work becomes more specific. As roads and walks appear, lighting may be designed to suit.

As always, where specific material is available it is incorporated into the estimate. Material of less precise description is estimated from the charts or, if still necessary for certain items, on a square-foot basis. At this stage manufacturers will be prepared to give a fairly close approximation of costs for miscellaneous systems.

Again costs are assembled in the cost model and the reporting form. Relevant project costs are added and adjusted.

4. *50% Document-Completion Estimate*

This estimate is based on design drawings that have progressed to midpoint. It has the advantage of making the design process visible to all parties. A certain amount of grief can be avoided by strict adherence to committed design schedules and cost estimates. Requiring formal recognition from a client before proceeding creates visible milestones.

By this stage the final cost begins to come into view. Adjustments should be made now. Design contingencies may be reduced, since many of the imponderables have disappeared.

Prepare the required cost model and reporting forms as before.

5. *100% Document Completion—Final Estimate*

When drawings are realistically past 90% completion and sufficient time yet remains for possible corrections before bid announcement, an accurate, takeoff-type estimate should be made. Prices of materials are obtained from vendors. Labor unions should be contacted to see if the wage rate used will continue through the life of the project or will be changed. Bid climate is reexamined to determine a propitious time. Contractor interest should be stimulated to enlarge the serious bidder list.

Now is the time to make necessary adjustments to suit budgetary requirements.

This final estimate is prepared with a cost model and submitted to the client.

In the period remaining before the bid date the design engineer should do his utmost to clarify his drawings. It should be noted that a set of documents gets bid before it is built. If for any reason something is unclear to the contractor's estimator or looks ponderous to him, the job will be penalized. Items perfectly clear in intent but not 100% delineated on the drawings will haunt the designer in the form of extras.

Period Cost Models

The cost model may be used as part of the formal cost-control system. For in-house use, cost models should stay with the project manager. As systems evolve or are changed, effects on the cost model can be immediately registered. This is the simplest and quickest way to note cost changes before a design group has spent a lot of time on wasted effort. More formally, the cost model may be made a part of the entire project control system established between the owner and the design office. Cost models should be presented as parts of the formal design progress reports. Where cost-raising changes that occur are due to design-team effects, they will have to be cleared up. Where changes or requests for changes are made by the owner's representatives, their effects on costs should be brought to the owner's attention.

With the cost-model technique the cost effects of program changes can clearly be demonstrated in the periodic reports to the client. Authorized cost changes to the program are also conveniently isolated so that the owner can visualize changes in the design fee structure. In a project of some size, the "owner" is in fact a large group. At best all of "his" parts have their own particular interest in the pieces of the project, and probably none of them has an interest in the whole thing. A surgeon may not care about a hospital's façade, but he has a strong interest in the operating-room lighting, the closed-circuit television, and the parking lot. Dietitians could not care less about operating-room lighting, but they do want adequate power and voltages available for all their equipment and they do care about the parking lot. (In fact, possibly the only item of common interest and importance to an entire hospital is the parking lot. This might perhaps be considered the most important part of a hospital.) The result of this taffy pull of self interest can be expensive for the design office. Project managers trying to please the disparate groups are headed for cost-control problems.

A process of review should be established and agreed upon by all parties interested in the design process. This may be the single most important event in seeing a project successfully and profitably completed in the design office. Cost models should be presented at the times agreed upon. The electrical cost model is only one of those required, since each trade must contribute its own cost model. These models should be taken together and reviewed as a group. Not all design firms have all the required A/E services, but one person familiar with all the designers involved should assemble the trade cost models into a project cost model. In the periodic presentation to the client, backup material is then available to describe and locate the cost position of the project at the time of presentation.

The importance of recognition and acceptance of the cost models by the client's representative cannot be overemphasized. In writing a contract for architect/engineer services, careful notice should be given to cost control by way of a cost-model requirement. The A/E is in control of his design options when he need not be subject to perfidious changes suggested by the client. If the contract is written to require strict budget conformance, those changes affecting price represent a scope change in the design contract and, therefore, should command a change in fee. Since changes invariably slow down the momentum of a project and often cause extra expense, we prefer to see these limited in number and extent. A client who can be shown cost effects directly on the cost model may be inhibited from making too frequent changes due only to fancy. The presence of the cost model with subtrade backups at the periodic review is a handy way to contain unduly imaginative clients.

Another use of the cost model can be made by the members of the design team where a project requires the efforts of more than one trade. Cost models of each trade are assembled into a project cost model. Where budget adjustments become necessary, the cost model for each trade can serve to make adjustable items more visible. In the overall picture, electrical work may be represented by a single cost block, as seen in the EPA cost models, or may be broken out into the four main categories of electrical work indicated in the cost-model sheet of Fig. 2-11. As in the EPA general cost model, actual costs at large variance with possible costs can be isolated and their components expanded for analysis.

Cost Models may be indicated as part of a design process flow chart, as in Fig. 2-12. A document such as this, presented to the client as part of the owner–engineer agreement, clearly defines for both parties when cost models may be expected. Extra work beyond the periodic estimate is minimal if the list form of Fig. 2-12 is used. A design process flow chart indicating the responsibilities of each party during the process serves the interest of both parties. Clear defini-

TYPICAL DESIGN PROCESS FLOW CHART

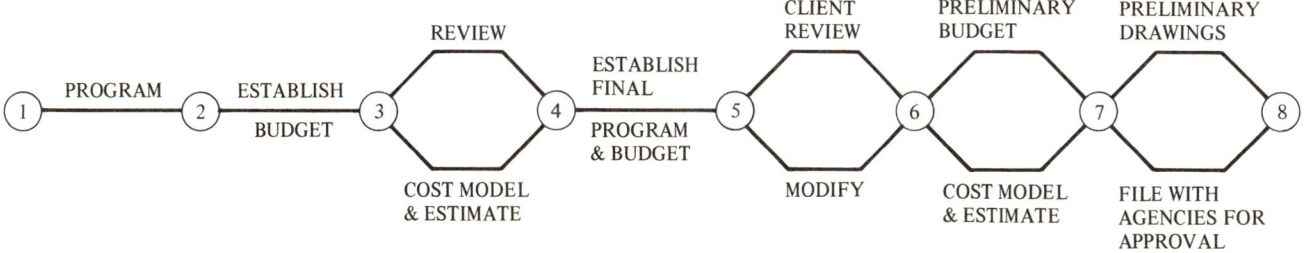

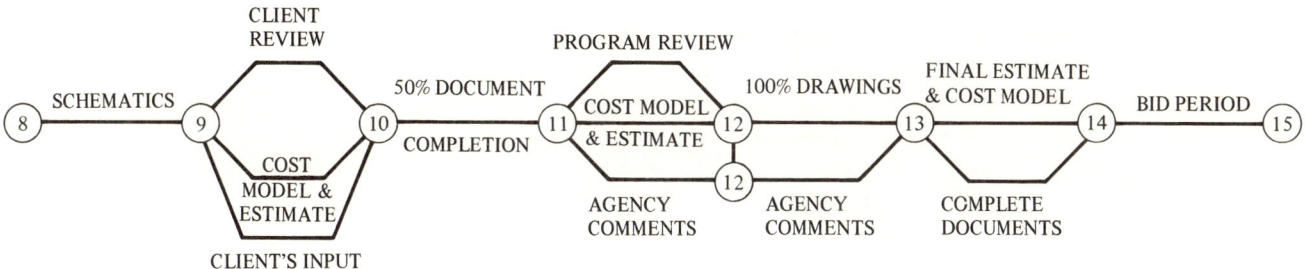

Fig. 2-12.

tion of action and reaction required by the parties to the design agreement reduces misunderstanding in the same way that good fences make good neighbors. Charts of this nature should either be drawn on a time scale or bear date flagmarks at certain milestones.

As suggested by the process flow chart of Fig. 2-12, reviews and cost summaries should be prepared at critical times. Estimates will be called:

a. Program
b. Budget—schematic phase
c. Preliminary—early design
d. 50% completion—50% documents
e. 100% completion—Pre-bid final drawings

It is convenient in the management of the design operation to present these estimates in a consistent form. All parties will be able to work with them and understand their significance.

Value Engineering

Value Engineering is a formalized technique for examining cost components in a construction project. This is yet another cost-control scheme. Generally speaking, the effects of Value Engineering are most dramatically seen in the more advanced stages of design. When solid commitments as to method have been made, a critical overview produces results that are obviously beneficial. The designer early in conceptual phases of a project is working with an amorphous group of concepts, determining what is right by a process of zeroing in to a workable solution. When a Value Engineer sees the job the hard work of major selections and process of elimination has already been completed. In contrast, the designer must of necessity make choices that only later may be subject to an intensive price probe, examining the pieces of the completed tapestry. His time and fee is largely spent in producing a workable solution out of a plethora of input information.

Value Engineering is a process of taking a job apart to examine its cost parts. A construction program is reviewed by groups of multi-disciplinary teams. A brainstorming session in these groups is used to generate alternative solutions to either the design or the program or both. The Value Engineer wants to evaluate worth, to produce a building function at minimum cost. Frequently, the operational process or basic concepts may be changed as a result of this kind of investigation. The result of brainstorming, then, is a list of ideas without regard to feasibility. Usually, the problem areas indicated by cost models as most fruitful subjects for economizing are the ones attacked.

An evaluation process follows the idea listing. Formal techniques of putting numerical values to design criteria allow the selection of two or three of the listed ideas as worthy of further exploration. These choices are roughly designed so that estimates can be prepared for each choice. Where the estimates do not produce results immediately comparable to each other, further analysis must follow. Basically, this step is necessary to put all costs on a common base, permitting cost comparison and economic choice. The process of life-cycle costing may also be used as a comparison method.

The Federal Government now requires a Value Engineering review for much of the architect/engineer contract work let by its agencies. Fees for this work are separate from the design fee. If the design is within the established budget, results of Value Engineering necessitating design changes should be paid for by the agency requesting the changes. Where there are apparent budgetary overruns, the design firm may have to bear the expense of document changes required to bring the project back on the money track.

Frequently, there is antipathy between the design team and the Value Engineers. Designers feel that their ideas and to some extent their professional competence is under attack. This is really not the case at all. A Value Engineer looks at a project from a completely different point of view from that of the designer. Value Engineers do not get involved in the dirty work of extracting a hard, workable program from groups or individuals, department heads, corporate officers, administrators, or other non-construction types of people. Generally, Value Engineers come as surgeons with cool eyes and steady hands to operate on somebody else's baby. But make no mis-

or at some time in the future, is placed on the common footing of present worth. This can then be readily compared to the initial cost of each system.

At this point, we would like to briefly explain the terms used in the life-cycle comparison. These explanations and illustrations are basic. All the expressions revolve about the compound interest formula.

(In order to perform the calculations in the sample life-cycle cost analysis problem at the end of this chapter, we have used values from the standard compound interest tables. The particular table of values for the interest rate in the problem and in the intervening examples is 10%. We have reproduced the table here for convenience as Fig. 2-13.)

If we were to invest a single sum at compound interest today, it would produce an amount equal to the principal plus an increasing annual amount of interest. This is, mathematically,

$$S = P(1 + i)^n$$

where S is the total amount after a number of periods, n is that number of periods. P is the principal invested, and i is the annual interest rate. Everybody is familiar with the effects of this function, since it is the arrangement used by banks in giving compound interest.

Example:

How much will $1,000 invested at 10% compounded annually amount to in 12 years?

$$S = ?$$
$$P = 1,000$$
$$i = 10\%$$
$$n = 12$$

$$S = P(1 + i)^n$$
$(1 + i)^n$ = compound amount factor (from the table)
$$= 3.1384$$

Therefore

$$S = \$1,000 \times 3.1384 = \$3,138.40$$

Present worth (PW) is a part of the compound interest expression. It is formally equivalent to the value P of the principal, the amount the interest works on to produce a new total. If we simply rearrange terms in the preceding expression so that

$$P = \frac{S}{(1 + i)^n}$$

we have an expression for present worth in terms of the final amount S. The table simply uses $S = 1$; then $P = 1/(1 + i)^n$. Conceptually, present worth represents that amount that would have to be invested now to produce the amount S after n periods at the interest rate i.

In the problem shown above, we can use $3,138.40 as an amount for which we would like to find the present worth.

	n	Amount of $1 at compound interest	Amount of an annuity of $1	Sinking fund to produce $1 in the future	Present value of $1	Present value of an annuity of $1	Annuity whose present value is $1	n	
	1	1.100 000 0000	1.000 000 0000	1.000 000 0000	.909 090 9091	.909 090 9091	1.100 000 0000	1	
	2	1.210 000 0000	2.100 000 0000	.476 190 4762	.826 446 2810	1.735 537 1901	.576 190 4762	2	
	3	1.331 000 0000	3.310 000 0000	.302 114 8036	.751 314 8009	2.486 851 9910	.402 114 8036	3	
	4	1.464 100 0000	4.641 000 0000	.215 470 8037	.683 013 4554	3.169 865 4463	.315 470 8037	4	
	5	1.610 510 0000	6.105 100 0000	.163 797 4808	.620 921 3231	3.790 786 7694	.263 797 4808	5	
	6	1.771 561 0000	7.715 610 0000	.129 607 3804	.564 473 9301	4.355 260 6995	.229 607 3804	6	
	7	1.948 717 1000	9.487 171 0000	.105 405 4997	.513 158 1182	4.868 418 8177	.205 405 4997	7	
	8	2.143 588 8100	11.435 888 1000	.087 444 0176	.466 507 3802	5.334 926 1979	.187 444 0176	8	
	9	2.357 947 6910	13.579 476 9100	.073 640 5391	.424 097 6184	5.757 023 8163	.173 640 5391	9	
	10	2.593 742 4601	15.937 424 6010	.062 745 3949	.385 543 2894	6.144 567 1057	.162 745 3949	10	
	11	2.853 116 7061	18.531 167 0611	.053 963 1420	.350 493 8995	6.495 061 0052	.153 963 1420	11	
	12	3.138 428 3767	21.384 283 7672	.046 763 3151	.318 630 8177	6.813 691 8229	.146 763 3151	12	
	13	3.452 271 2144	24.522 712 1439	.040 778 5238	.289 664 3797	7.103 356 2026	.140 778 5238	13	
	14	3.797 498 3358	27.974 983 3583	.035 746 2232	.263 331 2543	7.366 687 4569	.135 746 2232	14	
	15	4.177 248 1694	31.772 481 6942	.031 473 7769	.239 392 0494	7.606 079 5063	.131 473 7769	15	
	16	4.594 972 9864	35.949 729 8636	.027 816 6207	.217 629 1358	7.823 708 6421	.127 816 6207	16	
	17	5.054 470 2850	40.544 702 8499	.024 664 1344	.197 844 6689	8.021 553 3110	.124 664 1344	17	
	18	5.559 917 3135	45.599 173 1349	.021 930 2222	.179 858 7899	8.201 412 1009	.121 930 2222	18	
.1 per period	19	6.115 909 0448	51.159 090 4484	.019 546 8682	.163 507 9908	8.364 920 0917	.119 546 8682	19	.1 per period
	20	6.727 499 9493	57.274 999 4933	.017 459 6248	.148 643 6280	8.513 563 7198	.117 459 6248	20	
ANNUALLY If compounded *annually* nominal annual rate is	21	7.400 249 9443	64.002 499 4426	.015 624 3898	.135 130 5709	8.648 694 2907	.115 624 3898	21	ANNUALLY If compounded *annually* nominal annual rate is
	22	8.140 274 9387	71.402 749 3868	.014 005 0630	.122 845 9736	8.771 540 2643	.114 005 0630	22	
	23	8.954 302 4326	79.543 024 3255	.012 571 8127	.111 678 1578	8.883 218 4221	.112 571 8127	23	
	24	9.849 732 6758	88.497 326 7581	.011 299 7764	.101 525 5980	8.984 744 0201	.111 299 7764	24	
10%	25	10.834 705 9434	98.347 059 4339	.010 168 0722	.092 295 9982	9.077 040 0182	.110 168 0722	25	10%
	26	11.918 176 5377	109.181 765 3773	.009 159 0386	.083 905 4529	9.160 945 4711	.109 159 0386	26	
	27	13.109 994 1915	121.099 941 9150	.008 257 6423	.076 277 6844	9.237 223 1556	.108 257 6423	27	
SEMIANNUALLY If compounded *semiannually* nominal annual rate is	28	14.420 993 6106	134.209 936 1065	.007 451 0132	.069 343 3495	9.306 566 5051	.107 451 0132	28	SEMIANNUALLY If compounded *semiannually* nominal annual rate is
	29	15.863 092 9717	148.630 929 7171	.006 728 0747	.063 039 4086	9.369 605 9137	.106 728 0747	29	
	30	17.449 402 2689	164.494 022 6889	.006 079 2483	.057 308 5533	9.426 914 4670	.106 079 2483	30	
	31	19.194 342 4958	181.943 424 9578	.005 496 2140	.052 098 6848	9.479 013 1518	.105 496 2140	31	
20%	32	21.113 776 7454	201.137 767 4535	.004 971 7167	.047 362 4407	9.526 375 5926	.104 971 7167	32	20%
	33	23.225 154 4199	222.251 544 1989	.004 499 4063	.043 056 7643	9.569 432 3569	.104 499 4063	33	
	34	25.547 669 8619	245.476 698 6188	.004 073 7064	.039 142 5130	9.608 574 8699	.104 073 7064	34	
QUARTERLY If compounded *quarterly* nominal annual rate is	35	28.102 436 8481	271.024 368 4806	.003 689 7051	.035 584 1027	9.644 158 9726	.103 689 7051	35	QUARTERLY If compounded *quarterly* nominal annual rate is
	36	30.912 680 5329	299.126 805 3287	.003 343 0638	.032 349 1843	9.676 508 1569	.103 343 0638	36	
	37	34.003 948 5862	330.039 485 8616	.003 029 9405	.029 408 3494	9.705 916 5063	.103 029 9405	37	
	38	37.404 343 4448	364.043 434 4477	.002 746 9250	.026 734 8631	9.732 651 3694	.102 746 9250	38	
	39	41.144 777 7893	401.447 777 8925	.002 490 9840	.024 304 4210	9.756 955 7903	.102 490 9840	39	
40%	40	45.259 255 5682	442.592 555 6818	.002 259 4144	.022 094 9282	9.779 050 7185	.102 259 4144	40	40%
MONTHLY If compounded *monthly* nominal annual rate is	41	49.785 181 1250	487.851 811 2499	.002 049 8028	.020 086 2983	9.799 137 0168	.102 049 8028	41	MONTHLY If compounded *monthly* nominal annual rate is
	42	54.763 699 2375	537.636 992 3749	.001 859 9911	.018 260 2712	9.817 397 2880	.101 859 9911	42	
	43	60.240 069 1612	592.400 691 6124	.001 688 0466	.016 600 2465	9.833 997 5345	.101 688 0466	43	
	44	66.264 076 0774	652.640 760 7737	.001 532 2365	.015 091 1332	9.849 088 6678	.101 532 2365	44	
	45	72.890 483 6851	718.904 836 8510	.001 391 0047	.013 719 2120	9.862 807 8798	.101 391 0047	45	
120%	46	80.179 532 0536	791.795 320 5361	.001 262 9527	.012 472 0109	9.875 279 8907	.101 262 9527	46	120%
	47	88.197 485 2590	871.974 852 5897	.001 146 8221	.011 338 1918	9.886 618 0825	.101 146 8221	47	
	48	97.017 233 7849	960.172 337 8487	.001 041 4797	.010 307 4470	9.896 925 5295	.101 041 4797	48	
	49	106.718 957 1634	1057.189 571 6336	.000 945 9041	.009 370 4064	9.906 295 9359	.100 945 9041	49	
	50	117.390 852 8797	1163.908 528 7970	.000 859 1740	.008 518 5513	9.914 814 4872	.100 859 1740	50	
	51	129.129 938 1677	1281.299 381 6766	.000 780 4577	.007 744 1375	9.922 558 6247	.100 780 4577	51	
	52	142.042 931 9844	1410.429 319 8443	.000 709 0040	.007 040 1250	9.929 598 7498	.100 709 0040	52	
	53	156.247 225 1829	1552.472 251 8287	.000 644 1339	.006 400 1137	9.935 998 8634	.100 644 1339	53	
	54	171.871 947 7012	1708.719 477 0116	.000 585 2336	.005 818 2851	9.941 817 1486	.100 585 2336	54	
	55	189.059 142 4713	1880.591 424 7128	.000 531 7476	.005 289 3501	9.947 106 4987	.100 531 7476	55	
	56	207.965 056 7184	2069.650 567 1841	.000 483 1734	.004 808 5001	9.951 914 9988	.100 483 1734	56	
	57	228.761 562 3902	2277.615 623 9025	.000 439 0556	.004 371 3637	9.956 286 3626	.100 439 0556	57	
	58	251.637 718 6293	2506.377 186 2927	.000 398 9822	.003 973 9670	9.960 260 3296	.100 398 9822	58	
	59	276.801 490 4922	2758.014 904 9220	.000 362 5796	.003 612 6973	9.963 873 0269	.100 362 5796	59	
	60	304.481 639 5414	3034.816 395 4142	.000 329 5092	.003 284 2703	9.967 157 2972	.100 329 5092	60	

Fig. 2-13.

$$\frac{1}{(1+i)^n} = \text{present worth factor} = v^n$$
$$v^n = 0.31863 \text{ (from the table)}$$
$$P = \frac{S}{(1+i)^n} = Sv^n$$
$$P = \$3,138.40 \times 0.31863 = \$999.99 \text{ or } \$1,000.000$$

The present worth factor is used in a life-cycle analysis to bring any future one-shot payments back to the present to permit comparison of events which may occur at different times.

Example:

Which investment is worth more today?

  A yields $5,000 in 6 years at a present cost of $2,000.
  B yields $6,000 in 7 years at a present cost of $3,000.

If money can be invested at 10%, v^n (6 years) = 0.56447, and the present worth of A is

$$P = \$5,000 \times 0.56447 = \$2,822.35$$

Over 7 years, v^n = 0.51316, and the present worth of B is:

$$P = \$6,000 \times 0.51316 = \$3,078.96$$

Investment A has a present worth which is 1.411 times its present cost. Investment B has a present worth 1.026 times its present cost; therefore, investment A is the better choice.

Capital Recovery Costs

It is a usual requirement that inital construction funds borrowed be paid back to the lender. To do this an annuity or annual payment must be established. This payment consists of interest on principal outstanding during the previous year, plus partial repayment of principal. The annual payments represent the amount required to return the total interest plus capital at the end of the time period involved. This factor is called, variously, *annuity whose present value is 1* or *amortization factor* (*AF*), or *capital recovery factor*. The values of these payments may be found in stardard tables; mathematically, they are

$$AF = \frac{i(1+i)^n}{(1+i)^n - 1}$$

Example:

$$i = 10\% \quad n = 12 \text{ years}$$

Calculated value:

$$AF = \frac{0.10(1+0.10)^{12}}{(1+0.10)^{12} - 1} = 0.1467633151$$

From the tables:

$$AF = 0.1467633$$

The annual payments required for recovery of the initial cost and for recovery of one-shot costs together represent the annual expense for the use of borrowed capital. Each of the separate payments, whether for initial-cost or one-shot-cost recovery, are amortized over the entire life cycle. One-shot expenses are brought to the present and then amortized over the life cycle.

Example:

Interest rate: 10%;
Life cycle: 20 years;
One-shot expense: $5,000 in year 10.

Find the annual expense of amortization.

$$\text{annual expense} = \text{amount} (\$5,000) \times \text{PW factor} (v^n) \\ \times \text{amortization factor (AF)}$$

$$v^n \text{ (10 years)} = 0.38554$$
$$AF \text{ (20 years)} = 0.11746$$

$$\text{annual expense} = \$5,000 \times 0.38554 \times 0.1176 = \$226.43$$

This payment, $226.43, each year for the 20-year life cycle will be carried as an annual expense of purposes of comparison.

Salvage Value

Any recoverable value left in the property at the time of its replacement or completion of its life cycle should be included as a single expense item of negative effect. The annual reduction in expense is obtained in the same manner as outlined above. Savings realized will be in the form of the amortization of the present worth over the entire life cycle.

Annual Costs

Each construction alternative will have its own set of maintenance and operating costs. These may be critical in the establishment of the best alternative. Of these costs, that of energy is perhaps the most important, and energy estimates should receive the careful treatment they deserve. Maintenance costs will usually be a percentage of the initial cost if system alternatives use generally similar equipment. Unique maintenance requirements, of course, should be accounted for fairly. For example, certain sizes and types of refrigeration equipment require employment of skilled operators by code where other choices will not, and in comparing alternatives which differ in this regard payroll expenses have to be considered.

Comparison

As discussed earlier, the common base for comparison of alternatives is present worth. All the costs which have, in the preceding steps, been put on an annual basis over the life cycle of the facility are summarized as a total annual owning and operating cost. This total annual payment can be brought to the present by applying the PW factor for a uniform annual series. Mathematically this is

$$a_n = \frac{1 - v^n}{i} = \frac{1}{AF}$$

where $v^n = 1/(1 + i)^n$ as before

Example:

$i = 10\%$; $n = 15$ years

$$v^n = \frac{1}{(1 + 0.1)^{15}} = 0.239392$$
$$a_n = \frac{1 - .23392}{0.10} = 07.6060795$$

From the table, $a_n = 7.6060795$.

Example:

What is the present worth of energy payments of $1,000 per month over the 20-year life cycle of a project?
  Interest = 10%
  annual payment = 12 × 1000 = $12,000
  a_n (from table) = 8.513563
  PW = annual payment × a_n
      = $12,000 × 8.513563 = $102,162.76

Recapitulation

Life-cycle cost analyses must use a common base in order to make comparisons of financial events occurring at different times. This common base is the present worth (PW). All costs, initial and one-shot, are amortized over the full life cycle of the project. This annual payment plus the annual costs of maintenance and operation become a series of total annual payments for which the present worth is found. Comparing these for two or three alternatives shows which will represent the least cost.

Formulas may be used rather than interest tables. These formulas are:

Compound interest: What a dollar invested today will be worth in the future.
  S = sum, P = principal, i = interest rate;
  n = number of interest periods;
  $S = P(1 + i)^n$; when $P = 1$, $S = (1 + i)^n$.

Present worth: What a dollar at some time in the future is worth today.

$$PW = \text{present worth} = \frac{1}{(1+i)^n}.$$

Amortization factor/capital recovery factor/annuity whose present value is 1: The annual payment required to pay off $1 in *n* periods.

$$AF = \frac{i(1+i)^n}{(1+i)^n - 1}$$

Present worth of an annuity: Present value of a series of payments made each period over *n* periods.

$$a_n = \frac{(1+i)^n - 1}{i(1+i)^n}$$

Needless to say, in life-cycle analysis, as in other estimating efforts, the best data produce the best results. Since the cost of electrical energy will often be the single largest factor in the analysis, an exact statement of the local utility rate schedule must be obtained. Rates vary widely throughout the country, and sometimes will have a wide range in a small geographic area. Certain municipalities own their own electrical utility, and these rates may be lower than commercial rates in neighboring communities.

Example:

As an example, compare the cost of operating lighting in a space in New York City (Consolidated Edison Company of New York Inc.) with the same operation in Phoenix, Arizona (Arizona Public Service Company). Refer to copies of the rate schedule to follow the calculations (Figs. 2-14 and 2-15).

Load: 100 KW;
Hours of monthly use: 200;
Total KWH per month: 20,000;
Low-tension service.

New York City

Demand Charge: $8.79/KW × 100 KW	= $ 879.00
Energy Charge: $.0369/KWH × 20,000 KWH	= $ 378.00
	$1617.00

$$\text{Actual Cost/KWH} = \frac{\$1617}{20{,}000 \text{ KWH}} = \$ \ 0.0809$$

Phoenix

Cumulative KWH	KWH × Rate	Charges
12	12	$ 1.55
500	488 × 0.0505	24.64
3500	3000 × 0.0427	128.10
31155*	27655 × 0.0246	680.31
		834.60
	+11.81%	98.57
		$933.17

$$\text{Actual Cost/KWH} = \frac{\$933.17}{20{,}000 \text{ KWH}} = \$ \ 0.0467$$

*97 × 115 = 11,155 KWH adjustment for size of load added as per schedules.

ELECTRIC RATES

ARIZONA PUBLIC SERVICE COMPANY
Phoenix, Arizona
Filed by: B. Paul Hart
Title: Manager, Rate Services
Date Original Filing: April 6, 1931
District: Company's Rate Area No. 1
      As Specified Under Availability

A.C.C. No. 2210
Cancelling A.C.C. No. 2110
Tariff or Schedule No. E 32 1
Twentieth Revisal Sheet No. 1
Effective: April 23, 1976

Filed: April 23, 1976

GENERAL SERVICE

AVAILABILITY

In Company's Rate Area No. 1 (including but not limited to Buckeye, Chandler, Gilbert, Glendale, Goodyear, Litchfield Park, Peoria, Phoenix, Scottsdale, Sunnyslope, Tempe and environs) at all points where facilities of adequate capacity and the required phase and suitable voltage are adjacent to the premises served.

APPLICATION

To all electric service required when such service is supplied at one point of delivery and measured through one meter.

Not applicable to temporary, breakdown, standby, supplementary, or resale service.

TYPE OF SERVICE

Single or three phase, 60 Hertz, at one standard voltage (12,5000; 2400/4160; 480; 277/480; 120/240 or 120/208 volts as may be selected by customer subject to availability at the premises). Three phase service is furnished under Company's standard rules covering line extensions. Transformation equipment is included in cost of extension. Three phase service is not furnished for motors of an individual rated capacity of less than 7-1/2 HP, except for existing facilities or where total aggregate HP of all connected three phase motors exceeds 12 HP. Three phase service is required for motors of an individual rated capacity of more than 7-1/2 HP.

MONTHLY BILL**

RATE	$1.55 which includes the use of 12 kwh
	5.05¢ per kwh next 488* kwh
	4.27¢ per kwh next 3,000 kwh
	2.46¢ per kwh next 42,000 kwh
	2.20¢ per kwh next 400,000 kwh
	2.06¢ per kwh all additional kwh
	*Add 115 kwh for the first 197 kw over kw, 78 kwh per kw next 200 kw and 51 kwh per kw for all additional kw.
	Minus 0.135¢ for each kwh in excess of 300 kwh per kw or 12,000 kwh, whichever is the greater.
MINIMUM	$1.42 plus 86¢ for each kw in excess of 3 of the highest kw established during the 12 months ending with the current month, or the minimum kw specified in the Agreement for Service, whichever is the greater.
ADJUSTMENTS	(1) Subject to a purchased power and fuel (PPF) unit cost adjustment of plus or minus .0001¢/kwh for each .0001¢/kwh by which the PPF unit cost to the Company's electric operations exceeds or is less than 1.0135¢/kwh. The method of application is set forth in the filed "Plan for Administration of Adjustment for Purchased Power and Fuel Cost".
	(2) Plus the applicable proportionate part of any taxes or governmental impositions which are or may in the future be assessed on the basis of gross revenues of the Company and/or the price or revenue from the electric energy or service sold and/or the volume of energy generated or purchased for sale and/or sold hereunder.

DETERMINATION OF KW

The average kw supplied during the 15-minute period of maximum use during the month, as determined from readings of the Company's meter.

CONTRACT PERIOD

Up to 50 kw: As provided in Company's standard Agreement for Service
Over 50 kw: Three (3) years, or longer, at Company's option.

TERMS AND CONDITIONS

Subject to the Company's Terms and Conditions for the sale of electric service.

**In addition to the stated Rate and Minimum, 11.81% will be added to each bill, prior to the application of Adjustments (1) and (2).

Fig. 2-14.

SERVICE CLASSIFICATION No. 9

GENERAL–LARGE

Application to Use of Service for

Light, heat and power for general uses where the Customer's requirements are in excess of 10 kilowatts, subject to the Special Provisions hereof.

Character of Service

Of the various characteristics of service listed and more fully described in General Rule III-2, the following may be designated for service by the Company under this Service Classification, subject to the limitations set forth in such Rule. Frequencies and voltages shown are approximate. All are continuous.

Standard Service

Any derivative of the standard alternating current, 3 phase, 4 wire system at 60 cycles and 120/208 volts.

Non-Standard Service

Direct Current at 120, 120/240 or 240 volts.
Low Tension Alternating Current–60 cycles:
    Single phase at 120/240 volts
    Three phase at 265/460 volts
    Three phase at 240 volts
    Two phase at 120/240 volts or 230 or 240 volts

High Tension Alternating Current–60 cycles:
    Three phase at 2,400/4,150 volts
    Three phase at 3,000 or 7,800 volts
    Three phase at 6,900 volts
    Three phase at 13,200 volts
    Three phase at 33,000 volts
    Single phase and three phase at 2,400 volts

High Tension Alternating Current–25 cycles:
    Three phase at 6,600 volts

Rate

Demand Charge (per month)	Low Tension Service	High Tension Service
For the first 1,300 kw of maximum demand	$8.79 per kw	$8.46 per kw
For excess over 1,300 kw of maximum demand	$8.16 per kw	$7.83 per kw

There will be an additional charge of 60 cents per kilowatt of demand during the summer period. The summer period will be defined as the first scheduled monthly billing period ending after May 15 and for four successive monthly billing periods thereafter.

Energy Charge (per month)–for both low and high tension service

For the first 1,500,000 kwhr	3.69 cents per kwhr
For excess over 1,500,000 kwhr	3.56 cents per kwhr

Provided,

(a) When 360 hours' use of the maximum demand for the month is less than 3,600,000 kwhr, the first 3,600,000 kwhr shall be billed at the above rates and the excess over 3,600,000 kwhr shall be billed at 3.15 cents per kwhr.

(b) When 360 hours' use of the maximum demand for the month is more than 3,600,000 kwhr, all excess kwhr over 360 hours' use of the maximum demand shall be billed at 3.15 cents per kwhr.

Consolidated Edison Company of New York.
Date Effective: April 8, 1976.

Fig. 2-15.

Life Cycle Cost Analysis
Using Annual Owning and Operating Costs

Project	Item	Team No.
SAMPLE PROBLEM	LIGHTING	

Description of present and alternate designs	Present SURF MTD 1 × 4-2L40W	Alt. #1 REC. MTD 2 × 4-4L40W	Alt. #2 REC MERC 100W
Initial Costs			
1. Base Cost	21,200	14,300	50,800
2. Interface Costs			
a.	5,940	2,900	8,780
b.			
c.			
3. Other Costs[1]			
a.			
b.			
c.			
Total Initial Cost Impact (IC)	27,140	17,200	59,580
Initial Cost Savings			
Single Expenditures (Present Worth)			
Year 5-594 lamps × 3.90 × .6209	1438		
Year 10-594 lamps × 3.90 × .3855	893		
Year 15-594 lamps × 3.90 × .2394	555		
Year			
Year 5-580 lamps × 3.90 × .6209		1404	
Year 10-580 lamps × 3.90 × .3855		872	
Year 15-580 lamps × 3.90 × .2394		542	
Year 6-439 lamps × 19.25 × .5645			4770
Year 12-439 lamps × 19.25 × .3186			2692
Year 18-439 lamps × 19.25 × .1800			1521
Annual Owning & Operating Costs[2]			
1. Capital Recovery[3] ICX Amortization Factor (AF) where AF = .1175 based on 20 year life cycle at 10%	3189	2021	7001
Replacement Cost - Amortization			
a. Year (PW × AF)	(5) 169	(5) 165	(6) 560
b. Year	(10) 105	(10) 102	(12) 316
c. Year	(15) 65	(15) 64	(18) 179
d. Year			
2. Annual Costs[4]			
a. 28 KW × 4000 hrs × .04	4480		
b. 27 KW × 4000 hrs × .04		4320	
c. 53 KW × 4000 hrs × .04			8480
3. Total Annual Owning & Operating Costs Sum of items 1. and 2. above	8008	6672	16,536
Present Worth of Annual Payments	68177	56803	140781

1. Any other item affecting initial system should be included.
2. Where using future costs these should be escalated as reasonably as possible.
3. A very sophisticated analysis would assign an estimated interest rate for each of the distant events, since interest rates are subject to changes both up and down.
4. For the sake of simplicity here, a constant energy rate is used, but this too could be escalated if considered necessary.

Fig. 2-16.

Sample Problem: Life-Cycle Analysis

For purposes of illustration here, we have chosen the lighting system of a project to be analyzed. Lighting represents a substantial part of the electrical work in a building and there are always alternative design schemes available. We can compare costs on a reduced foot-candle level, for example. Assume the initial design calls for a 100-foot-candle lighting level. We can compare this with life-cycle costs for operation at 75 and 50 foot-candles. The result would of course show both initial and life-cycle savings for the lower lighting intensity. This can be used to evaluate the necessity or desirability of a selected lighting level on an equal base. Another comparison could be made by using a constant foot-candle requirement and fixtures of equal efficiency but varying quality. Here the lower initial cost could be offset by increased maintenance and replacement costs. A third type of comparison could be made by considering lighting fixtures necessary to produce the same lighting level, but using fixtures with different efficiencies. This case will be the more subtle, since greater efficiency is usually bought at higher initial cost.

Problem:

Area to be lighted: 10,000 square feet.
Illumination required: 100 foot-candles.
Alternatives: 1. Surface-mounted fixtures using two 40-watt fluorescent lamps.
              2. Recessed 2' × 4' fixtures using four 40-watt fluorescent lamps.
              3. Recessed 11" round fixture using one 100-watt coated mercury lamp.
Operating hours per year: 4000.
Electrical energy cost: $0.04/KWH.
Life cycle: 20 years.
Ballast life: 4 years.
Lamplife—fluorescent: 20,000 hours.
Lamplife—mercury: 24,000 hours.
Interest rate: 10% for life of project.
All fixtures use 20-year life.
Lamp-replacement cost—fluorescent: $1.90 materials and $2.00 labor.
Lamp-replacement cost—mercury: $14.25 materials and $5.00 labor.

This material is now collated and entered in the appropriate spaces on the cost analysis sheet (Fig. 2-16). This illustrative form is set up for comparing three alternative installations, but if need be could be extended to compare any number of choices.

Conclusion:

This analysis shows a minimum cost for method 2 over the life cycle of the installation. In this case it is coincidental that the method with the lowest initial cost also proved to be the least expensive over the life of the installation. There will be times when the results would show that another choice, perhaps higher or lower in initial cost, would produce a lower life-cycle cost. This, of course, can occur because of the different expenses developing for each method over the life cycle under consideration.

3.
Material and Labor

SUMMARY

Labor and material prices vary from place to place around the country and are also subject to timely changes. This is why temporal and geographical adjustments must be made to any pricing system. A method to adjust prices obtained from the price charts to any materials or labor rates has been devised. The method considers the labor intensity in each line item of cost and also a materials cost index for the same items. The information is combined to produce a total cost factor which permits regional adjustments in costs to be handled simply.

Manpower requirements depend, basically on the specific nature of the work, but are modified by the particular conditions on each job. Where scheduled time or labor conditions are such that long hours of overtime are required for timely completion of the work, the effects on labor productivity are not insignificant. These are investigated to determine an effective wage rate.

Another of the uses to which estimates are put is in the establishment or checking of payment schedules. This serves as a means of fairly establishing the manner in which a contractor will get paid, and establishes the owner's cash flow for the project as well.

WAGE RATES

Electrical construction is a labor-directed cost. The percentage of labor in the total cost is larger than in equipment-oriented trades or industries. In the Northeast, with union labor hovering around the $20.00-per-hour mark, labor costs average 40–60% of the electrical contract. To aid in identifying costs and to permit a simple adjustment process for other than the base geographical location, each item in the cost charts of Chapter 5 bears an indication of the labor intensity factor (LIF). Labor costs vary widely from place to place, and since they radically affect the price, the cost adjustments should be made. Contractor-type estimates or more fully detailed labor and materials estimates will of course have man-hours tabulated separately and directly charged at the local labor rate.

The proper wage rate is an important selection. Rates used in the basic charts do not include overhead and profit but are payroll costs of the employer. The wage rate is not simply the amount paid to the electrical worker, but is the rate charged to the owner before the addition of the contractor's overhead and profit. Proper costs to use should be obtained from the local union or labor council, or from a reliable electrical contractor. Fringe benefits vary from locale to locale and should be verified by local trade representatives.

An example of a union wage-rate calculation is given below:

Electrician Wage Rate

10.67	base rate
1.280	12% vacation
2.027	19% national pension fund
.624	5.85% social security
.705	welfare fund
.50	pension fund
.10	industry fund
15.906	

say $16.00/hour

This then is the contractor's cost to keep an electrician on the job and is the figure the estimator would use in preparing his estimate.

Productivity

Figures in the estimating charts represent normal productivity. Included in these costs are; supervision by the foreman, layout work, materials handling, setup, measuring and cutting, and associated cleanup. Productivity is subject to some obvious and some subtle pressures. For example, a very large job may develop a lower production rate by difficulty in getting to the work areas and in communicating with and locating personnel. People can get lost. The economic climate has an effect. Some hard-pressed geographical areas are showing increased output per man-hour, probably because of the increased manpower pool available.

Each job will have a more or less constant productivity factor during the life of the job, with some tendancy for the tail-end work to languish. Productivity, as a job factor, may be calculated in a manner suggested by the National Electrical Contractors Association (NECA). They have a device called a job factor check sheet (Fig. 3-1) from which a factor is calculated and applied to the labor portion of the project cost. This then adjusts normal productivity to the peculiarities of a particular job. These factors must be given fair evaluation for the actual determination of contract cost and manning requirements. It should be recognized that unit prices and square-foot costs used in this book are specific for an average building in each building type. Job modifications should be applied to variations within building type.

The application of this productivity factor is for the purpose of determining the effective wage rate. This can be applied directly when using total-cost-factor charts. Basic charts in this book are based on so-called average conditions. Changes in these conditions result in effectively producing a different wage rate. It is obvious

DATE _____   JOB OR ESTIMATE NO. _____
                                                   JOB NAME _____

VARIABLE FACTORS	DEGREE 1	MAX. %	DEGREE 2	MAX. %	DEGREE 3	MAX. %	DEGREE 4	MAX. %	DEGREE 5	MAX. %	% ASSIGNED
TYPE OF BUILDING		-5		5		AV. 10		20		30	
1. Construction	Standard	0		1		2		4	Special	6	
2. Design	Simple	-1		1		2		4	Elaborate	6	
3. Floor Plan	Uniform	0		1		2		4	Complex	5	
4. Occupancy–Use	Usual	-1		0		1		2	Special	3	
5. Size–Floor Area	Small	-1		0		1		1	Large	2	
6. Extent of Electrical System	High Density	0		1		1		2	Low Density	3	
7. Design of Electrical System	Simple	-1		1		1		2	Complex	3	
8. Quality of Electrical Layout	Excellent	-1		0		0		1	Poor	2	
WORKING CONDITIONS		-5		3		10		20		30	
1. Location of Job	Close In	-1		0		1		2	Out of Town	3	
2. Weather Conditions	Excellent	0		1		2		4	Extremely Bad	6	
3. Working Space	Large–Clear	-1		0		1		2	Small-Cluttered	3	
4. Amt. Work by Other Trades	Very Little	-1		0		1		2	Very Congested	3	
5. Material Storage	Excellent	-1		0		1		2	None	3	
6. Shop & Bench Space	Excellent	0		1		2		4	None	6	
7. Material Hoisting Conditions	Excellent	-1		1		2		4	None	6	
8. Other											
9.											
GENERAL CONTRACTOR		-5		5		10		20		30	
1. Experience with Work	Considerable	-1		1		2		4	None	6	
2. Progress Maintained	Excellent	-2		2		4		8	Poor	12	
3. Coordination of Trades	Excellent	-2		2		4		8	Poor	12	
4. Other											
5.											
ELECTRICAL CONTRACTOR		-5		5		10		20		30	
1. Experience with Work	Considerable	-1		1		2		4	None	6	
2. Experienced Supervision	Available	-2		2		4		8	None	12	
3. Experienced Workers	100% Available	-1		1		2		4	None	6	
4. Adequate Tools & Equip.	Adequate	-1		1		2		4	None	6	
5. Other											

Add 1% For Each of _____ Floors

Total Job Factor _____ %

Fig. 3-1.

MATERIAL AND LABOR 69

that if work is to be done in a particularly bad climate, say where half the day may be lost because of the weather, the effective wage rate would double. To get the effective rate, then, the actual rate should be multiplied by the job factor. By this operation, a higher hourly rate may be used in determining the total cost factor where the job factor is high, and conversely a lower wage rate will apply where the job factor is low. Stated mathematically:

effective wage rate = hourly rate × job factor

For example:

hourly rate = $16.00
job factor = 1.4
effective wage rate = $16.00 × 1.4 = $22.40

This rate, $22.40, should be used for all estimates during the design of the project, unless of course, a rate change occurs. In this event, the new hourly rate should be multiplied by the same job factor to obtain the effective wage rate.

Labor Intensity Factor (LIF)

In general, each line item comprising the estimate consists of both material and labor costs. The cost of labor divided by the total line-item cost is what we have defined as the labor intensity factor (LIF). Naturally this ratio depends on the local cost of labor, since the number of hours required for a task is considered uniform for the large urban areas around the country. (Productivity is somewhat dependent on climate and type of job, as previously discussed.) LIF values are listed with each line item in the estimate form and on the charts for application by the user. Base LIFs have been established using 1976 material prices prevalent in the Northwest and labor at $20.00 per hour cost on the job.

If the user discovers that on a particular project, the anticipated productivity will vary greatly from the national standard, then he may apply additional multipliers. One type of multiplier is the job factor obtained from the chart previously described (Fig. 3-1). This will in effect change the local cost of labor by the ratio of productivity locally to that of the national standard. If the project merely has a total difficulty factor, then this can be applied to the bottom line rather than on a line-by-line basis. Items of work seriously at variance with the norm should have a factor applied to the particular line item they affect. For example, if high-voltage techniques are required in a locale where technicians are scarce, the relevant line item will be affected by an actual reduction in productivity. This has the effect of raising the unit cost of labor for that particular line item.

Materials Cost Index (MCI)

Materials cost index (MCI) also uses the New York City area as the base, and of course the base MCI is taken at 1.0. Various publications such as *Engineering News Record* publish material indexes which can be used to find a suitable MCI for other than the base geographical location. Not only can the cost factor charts be used to suit local materials price markets, but they can be used to adjust prices that change over a period of time. For this adjustment consider mid-1976 as the price base.

Materials prices do vary from place to place, but less drastically than labor rates. In the preparation of contractor-type estimates this difference is handled by asking for quotations on major equipment every time. Contractors will also subscribe to one pricing service or another to keep their files up to date. The design office preparing final estimates must also follow these procedures to update their cost files.

In the early design-phase estimates for which the charts of Chapter 5 are especially useful, materials prices should be adjusted at least annually, and more frequently for those specific items that become very active in the price market.

Total Cost Factor (TCF)

Factors affecting total line-item cost are productivity, labor rate, labor intensity factor, and the material-cost index. For the convenience of users of this book, a family of curves (Figs. 3-2 through 3-11) has been prepared combining these variables to yield the single multiplying factor called the total cost factor (TCF). The estimator in any geographical location or price climate can enter a curve using the adjusted local rate and materials cost index and extract a single multiplier. By use of the TCF the estimating charts in this book should be usable for some time in the future as well, since the TCF charts accept a wide range of rates.

Of necessity, the pricing charts in Chapter 5 use a single geographical location as a labor and materials base cost. In this case, the cost of labor has been taken at $20.00 per hour. This is the approximate cost to an electrical contractor to keep a union journeyman on the job in New York City. Costs include the employee wages, vacations, benefits, etc., which must be paid by the employer in accordance with union contracts. Materials cost index is taken as 1.0 based on 1976 prices prevalent in the Northeast.

TCF charts plot a family of curves for each labor rate. Curves plot the labor intensity factor, against the total cost factor for each of several materials cost indices. To use a chart for adjusting estimated costs follow the following steps:

1. Select TCF chart for proper labor rate.
2. Use LIF from cost chart of Chapter 5.
3. Enter chart at LIF. Move horizontally to proper MCI and read TCF vertically below this point.

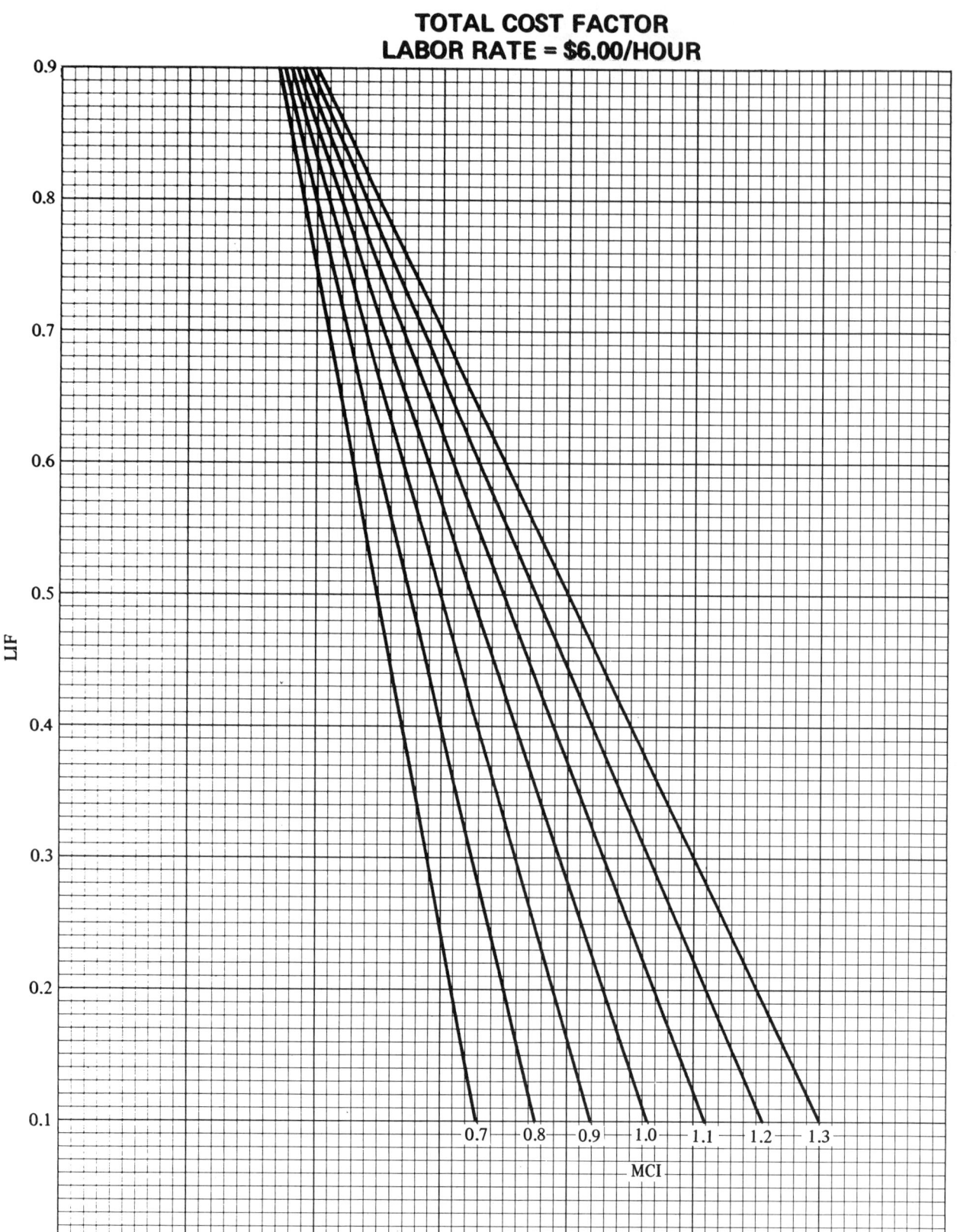

Fig. 3-2.

72 ESTIMATING AND COST CONTROL IN ELECTRICAL CONSTRUCTION DESIGN

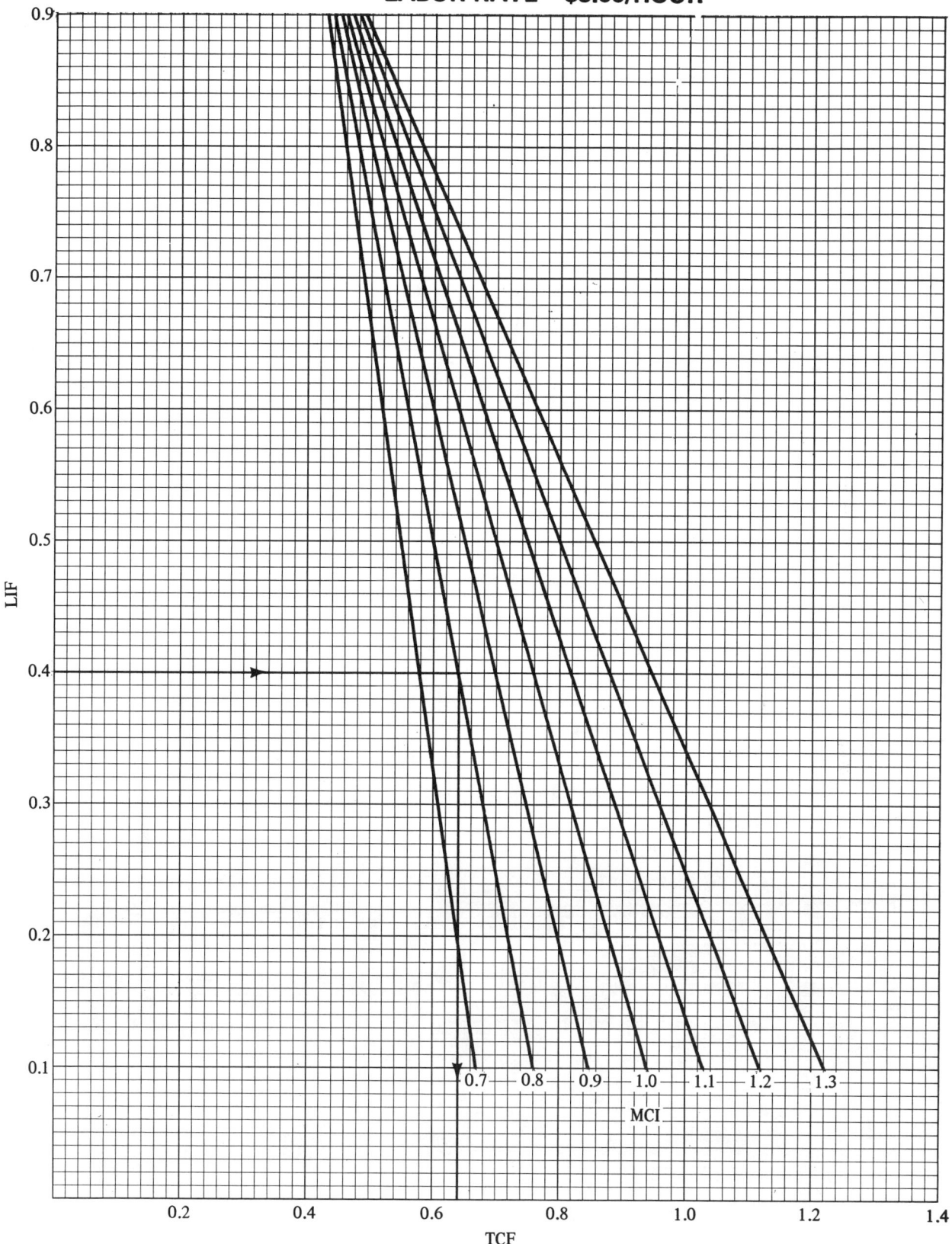

Fig. 3-3.

MATERIAL AND LABOR 73

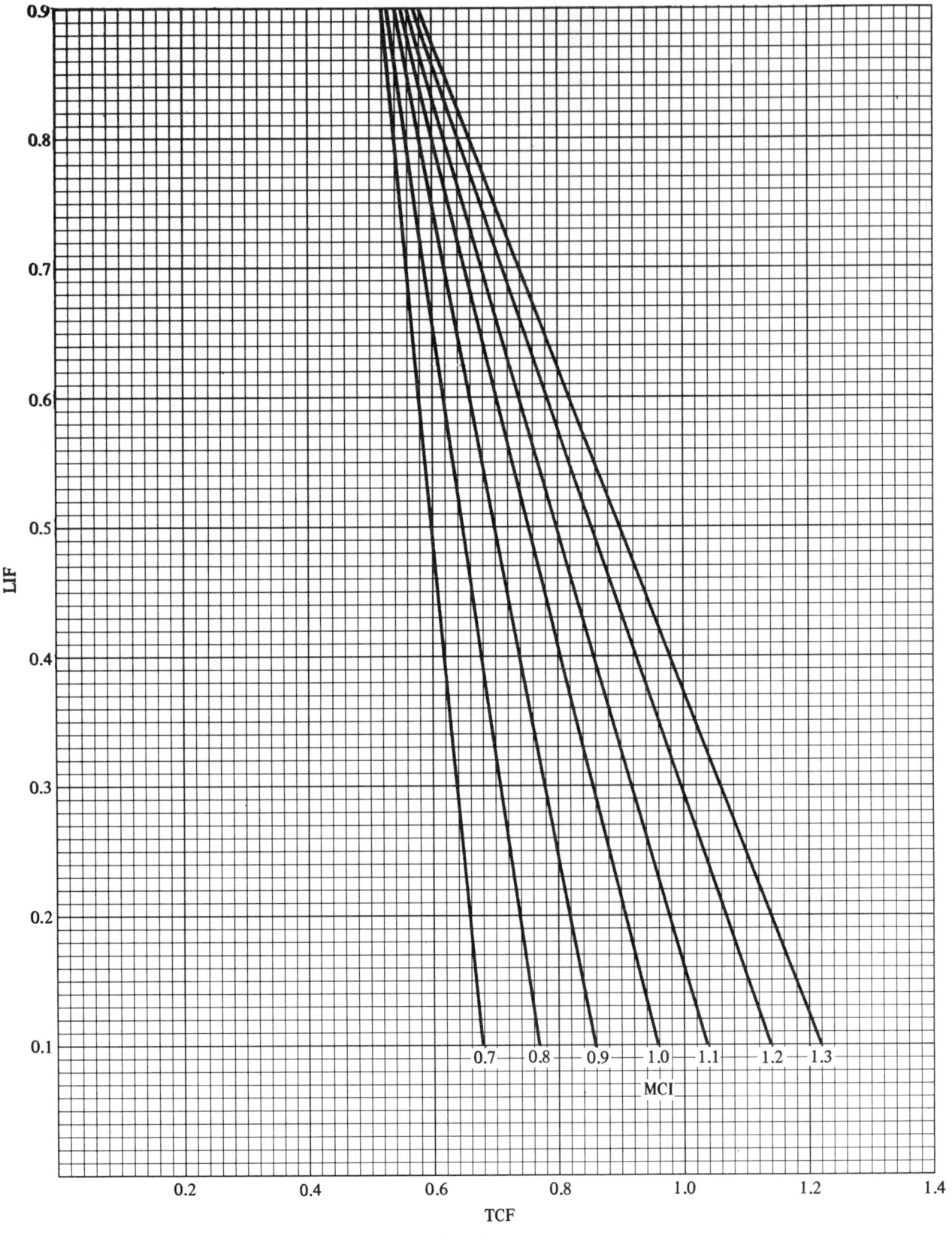

Fig. 3-4.

74 ESTIMATING AND COST CONTROL IN ELECTRICAL CONSTRUCTION DESIGN

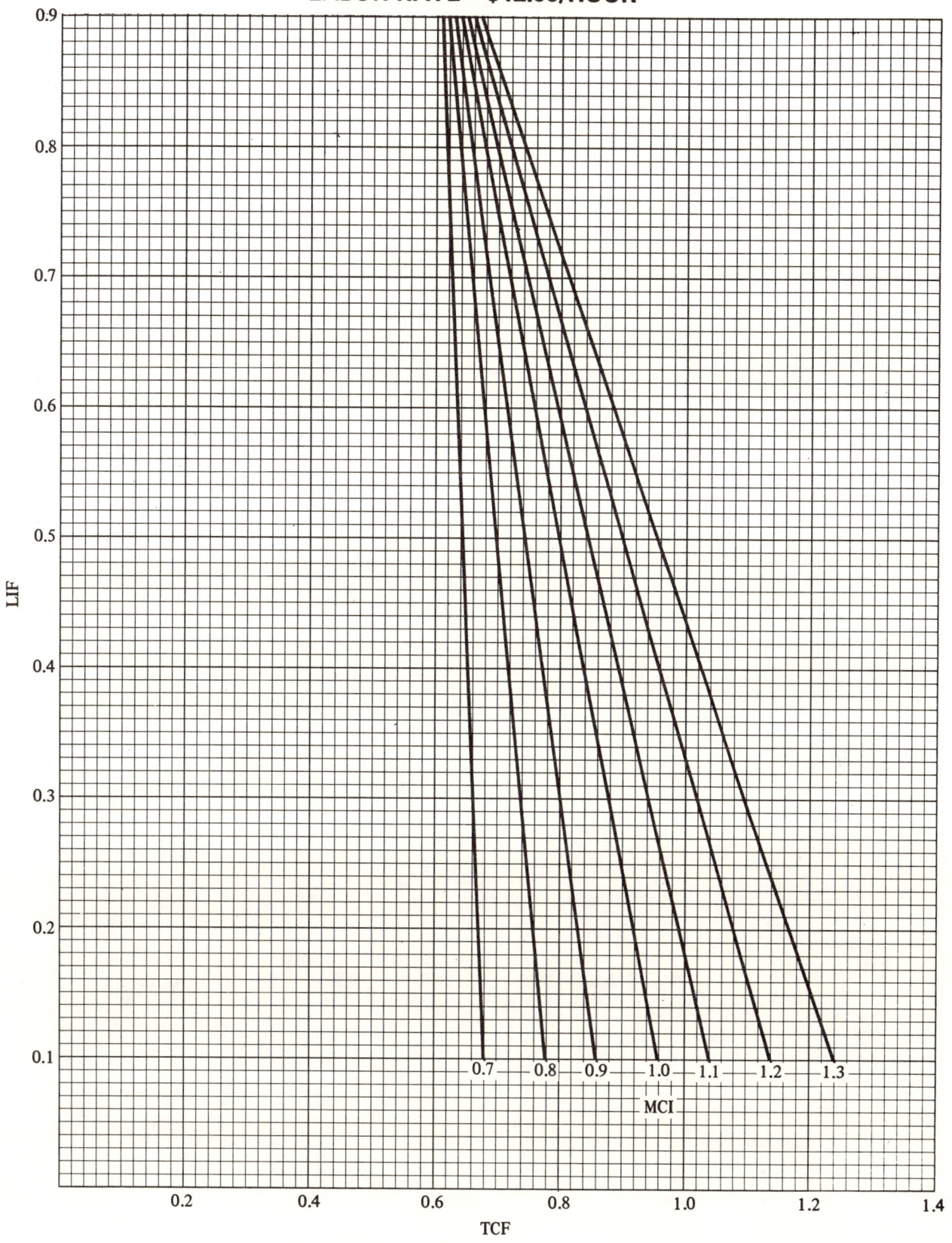

Fig. 3-5.

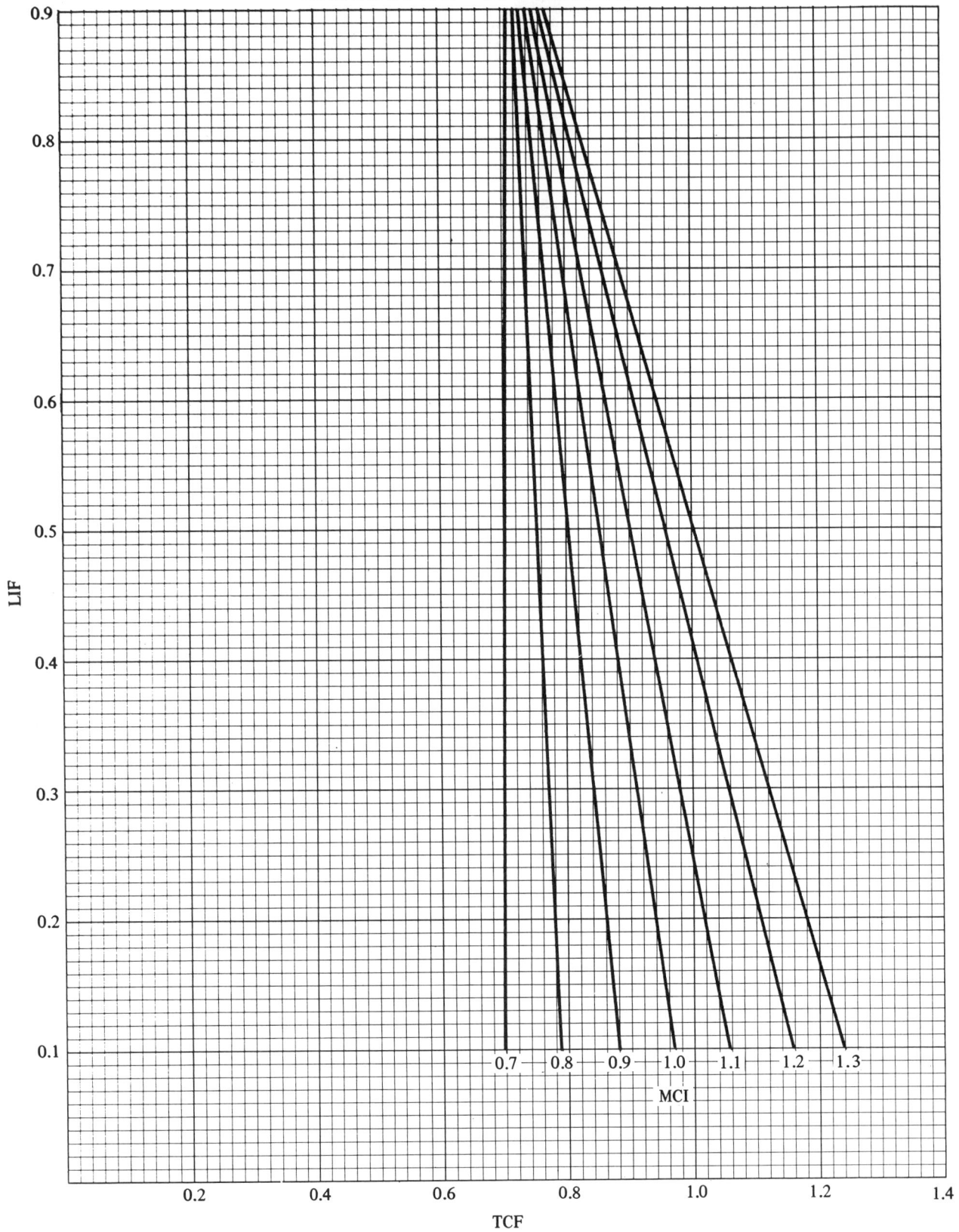

Fig. 3-6.

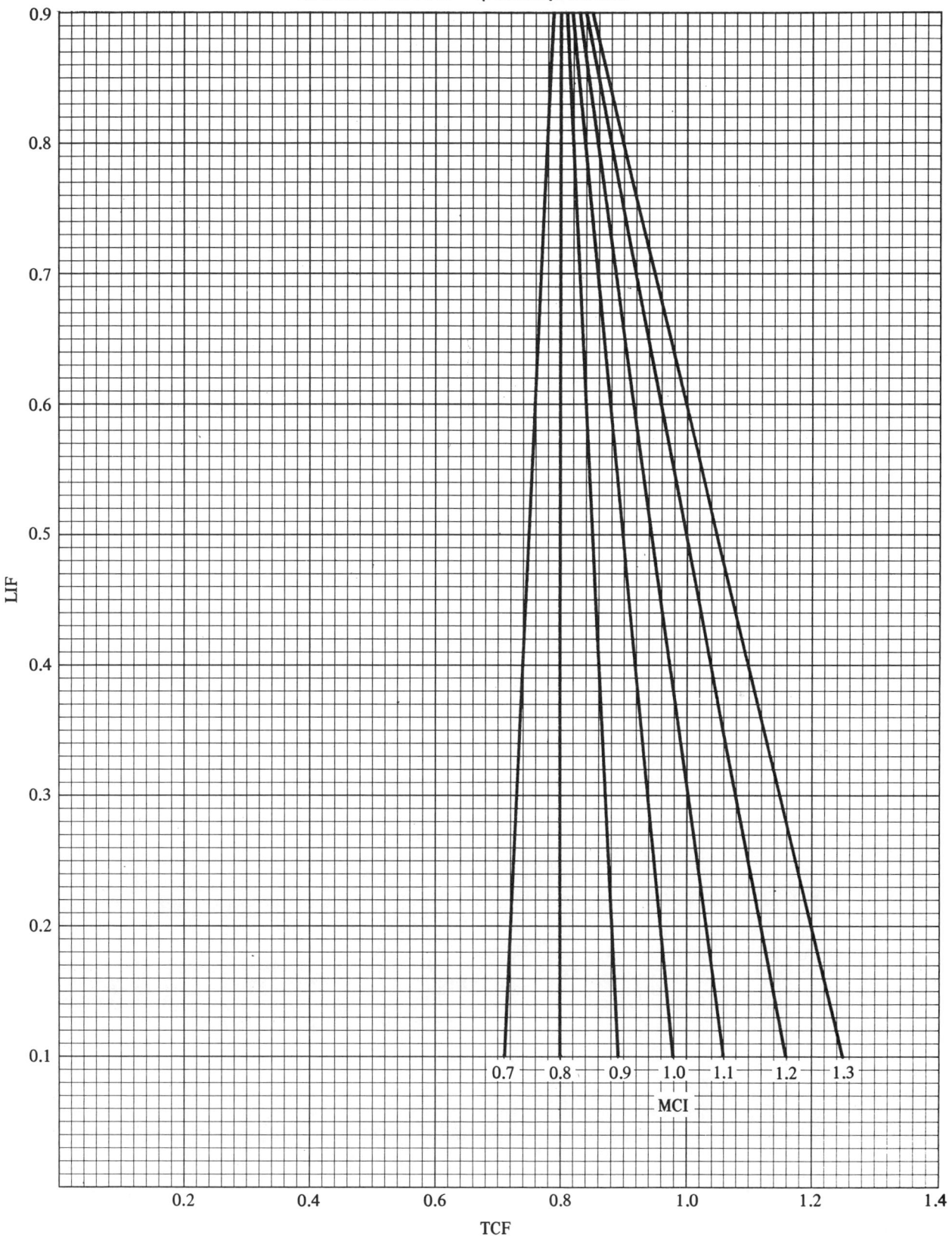

Fig. 3-7.

MATERIAL AND LABOR 77

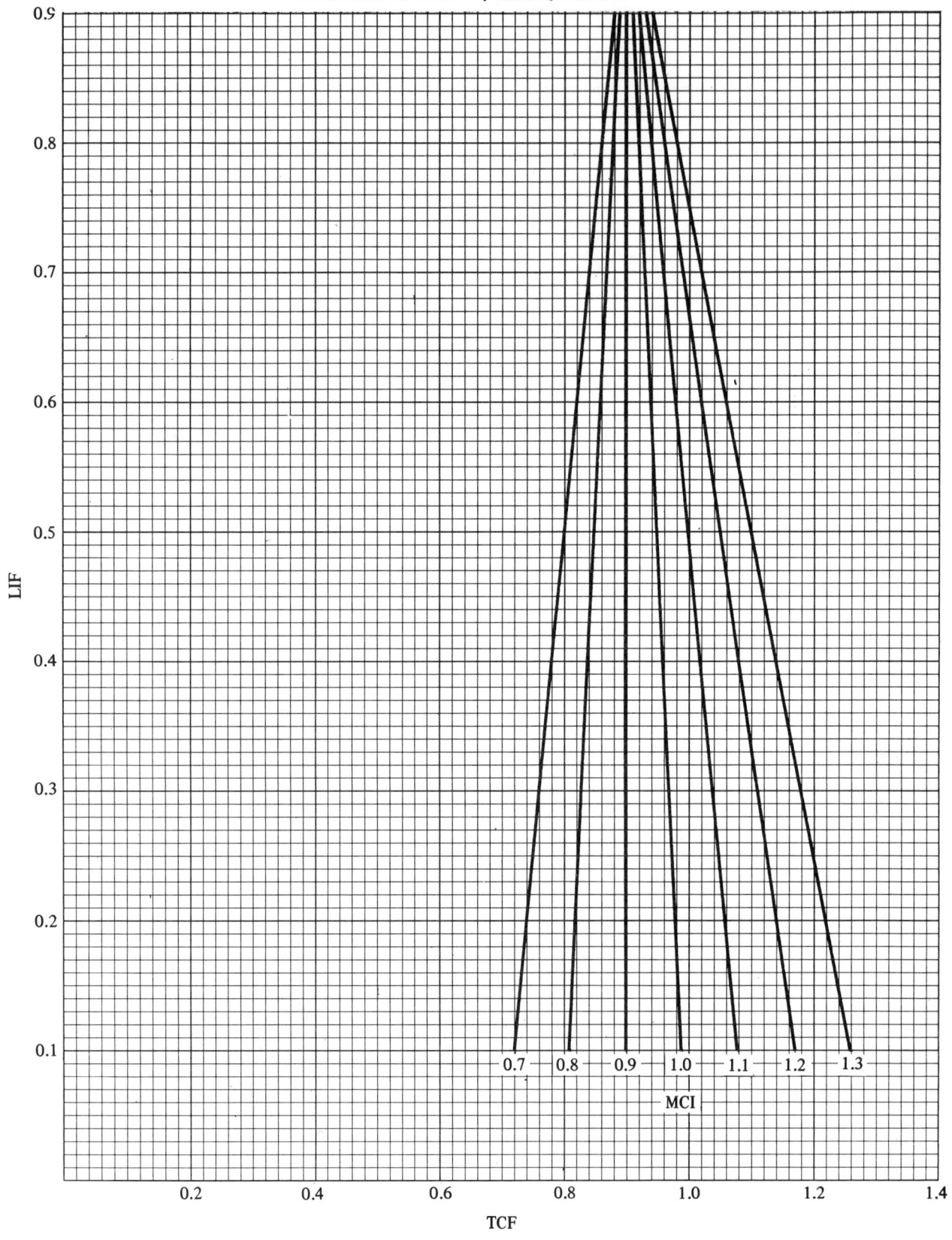

Fig. 3-8.

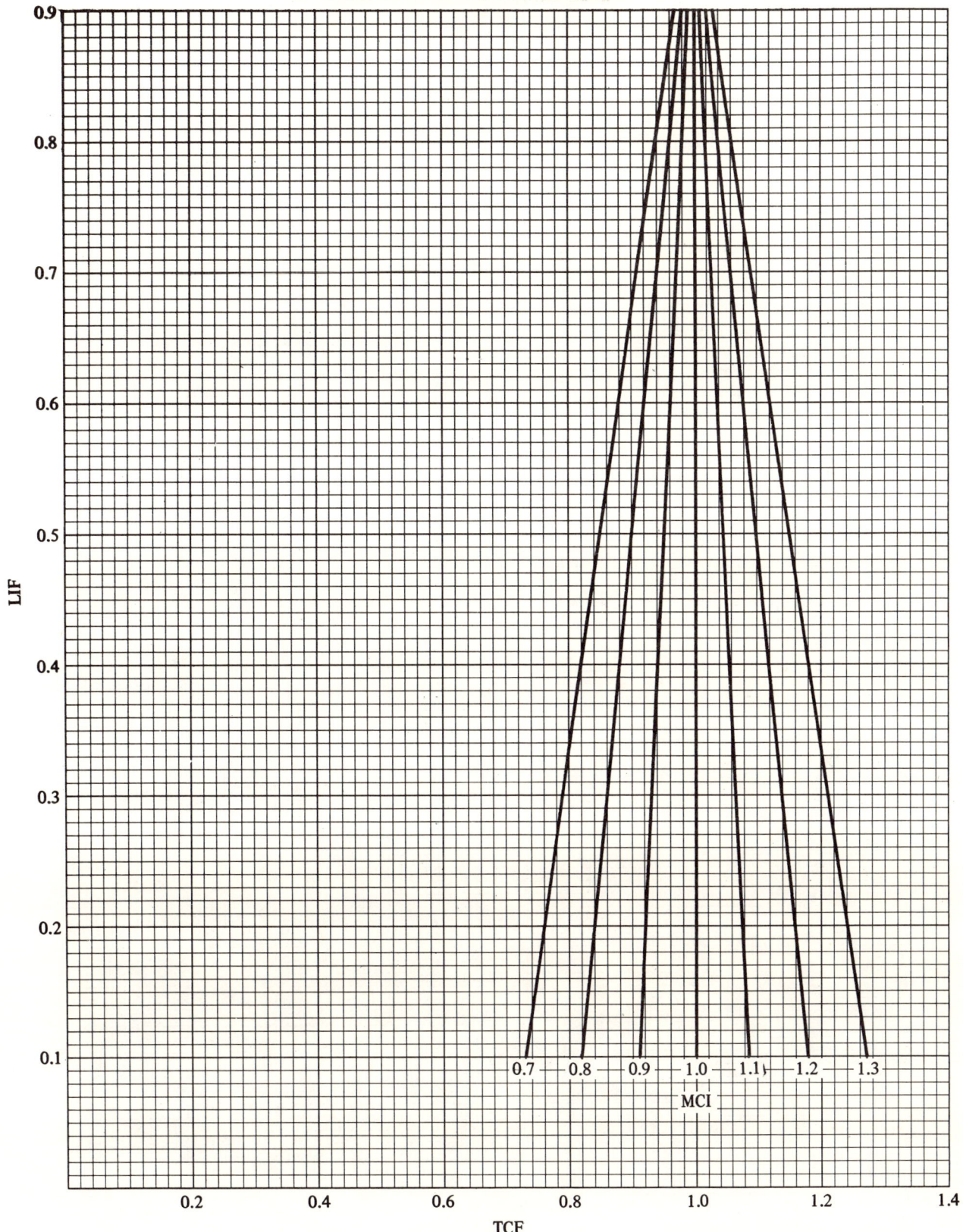

Fig. 3-9.

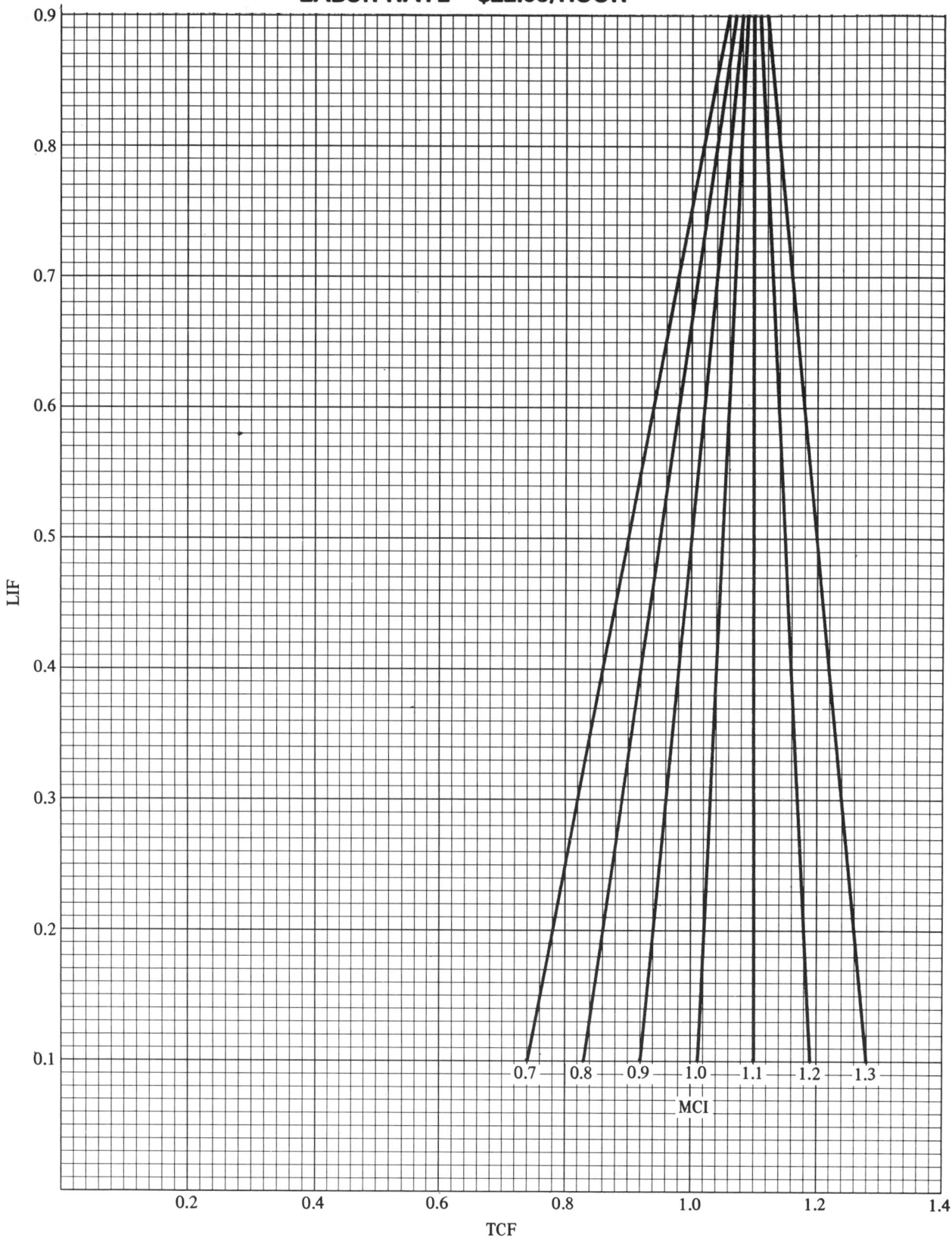

Fig. 3-10.

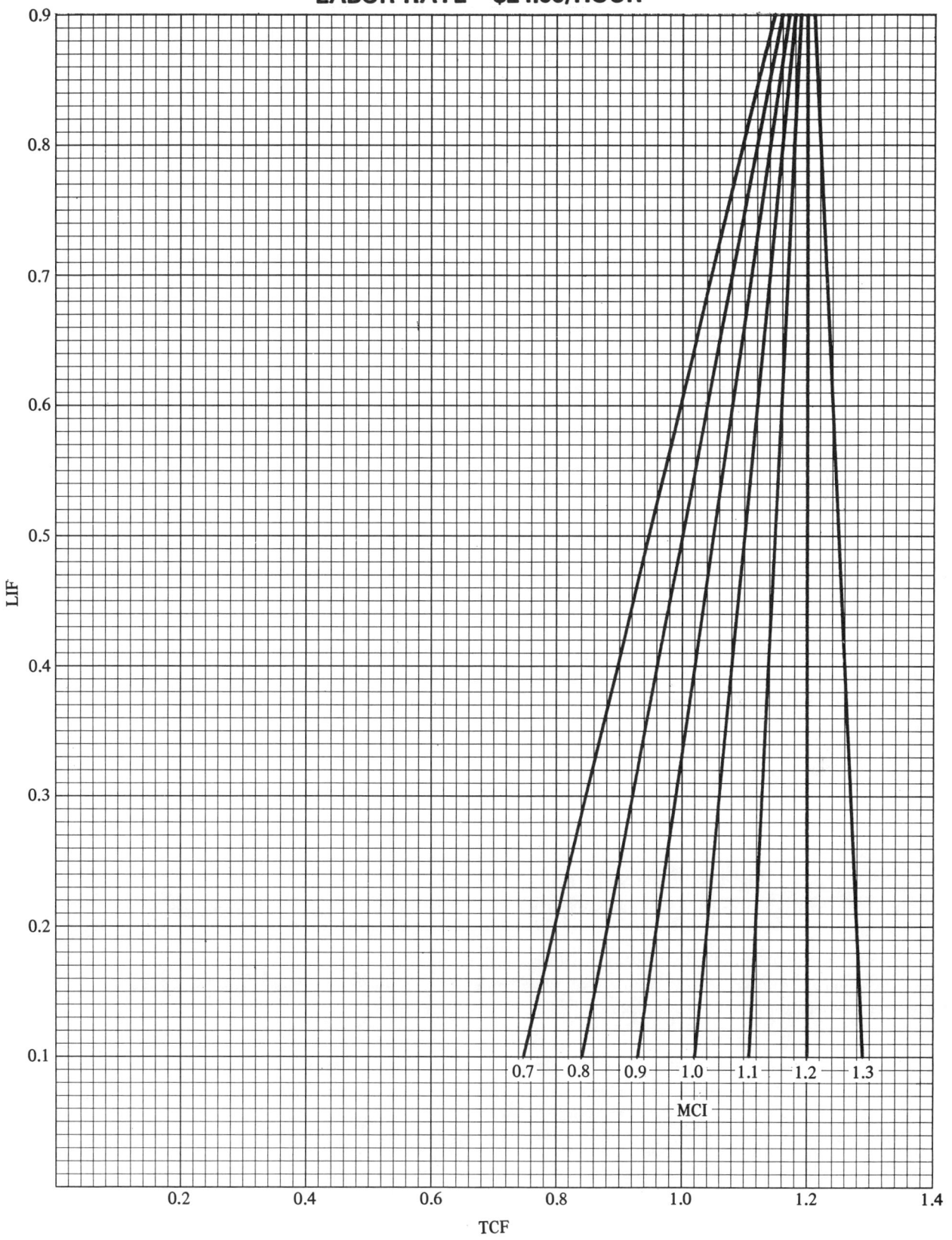

Fig. 3-11.

MATERIAL AND LABOR 81

Example:

    labor rate = $8.00/hour
    LIF = 0.4
    MCI = 0.8
    Therefore, TCF = 0.64

This example is illustrated on Fig. 3-3.

$$\text{local cost} = \text{base cost} \times \text{TCF}$$

This indicates that the particular item work at the location selected will be installed for approximately 64% of the cost of that at the base location (labor rate, $20.00/Hour; MCI = 1).

Since labor intensity varies with each item of work, adjustments to cost should be made on a line-item basis. For this reason, each of the cost charts in Chapter 5 indicates the applicable LIF.

Using the LIF and MCI, the TCF for each line item can readily be found. Once the user establishes a labor rate and MCI he will use repeatedly, he may for convenience enter the TCF directly on the cost charts. Values to be used in assembling the estimate may then be taken off directly if this is found more convenient.

MANNING REQUIREMENTS

One of the by-products of the estimating effort is the quantifying of labor involved in a project. In order for construction schedules to be met, each of the component members of the design team should be able to input data on construction time. Preliminary bar graphs or rough CPM drawings must be made early in the design process to determine feasibility of proposed construction time. Should these show an impossible condition, some of the parameters will have to be flexed. The earlier this occurs, the smaller will be the damage to a set of documents and/or the owner's plans. It is up to members of the design team to observe and prevent the establishment or development of a schedule that cannot be realistically met.

Schedules will first be drawn to show the gross features of a project and then refined to where daily or weekly development will be readily visible. This effort can parallel that of the estimating effort. Early rough estimates will produce rough estimates of man-hours required. Further refinement of plans and consequently of the later estimates will also allow more specific labor duration schedules. Charts in this book indicate the labor content of each of the tasks. This information appears in the form of the LIF on each of the charts. Labor figures developed from these data can be assembled in any form required by the scheduler. In this manner, labor hours required for electrical work events parallelling general construction events can be determined. The hours required are then used to calculate required crew days to complete any particular phase of work in a scheduled time. Should crew size become excessive, the scheduler should be so advised. One thousand men working a day may represent one thousand man-days, but the output may not equal, say, that of a 50-man crew working 20 days.

To determine man-hours in a task, use the following procedure:

1. Determine total cost from chart information.
2. Note LIF from chart.
3. Divide product of 1 and 2 above by 20.
4. The result is the number of man-hours.

Example:

Area = 100,000 square feet.
Hospital Building.
Lighting Panels, Chart 1C1.
208/120 volt.

$$\text{cost} = 0.16/\text{sq. ft} \times 100,000 = \$16,000$$
$$\text{LIF} = 0.63$$
$$\text{man-hours} = \frac{\text{cost} \times \text{LIF}}{20}$$
$$= \frac{16,000 \times 0.63}{20} = 504 \text{ hours}$$

Now, if two men are assigned to this work, the scheduler will have to be informed that the task will take

$$\frac{504 \text{ man-hours}}{2 \text{ men} \times 8 \text{ hours/day}} = 31.5 \text{ work days}$$

If this is not compatible with the precedence network, crews will have to be augmented to suit. The alternative, that of having the men work faster, is not usually employed.

In like manner, other lines in the estimate may be analyzed for labor content. Groupings of any required composition may then be made. If a project consists of several buildings, labor may be analyzed by building should the CPM schedule so require. As with so much work in recent times, there are heavy legal overtones to the plausibility of schedules. Many construction contracts are now negotiated wherein a rough CPM network forms part of the plan-specification body. The network is intended to demonstrate that the project described in the bidding documents can be produced within the allotted time. Should this prove not to be the case during the actual construction phase, the scheduler and possibly the design engineer who provided schedule information become exposed to claim actions.

Figure 3-12 is a convenient chart for estimating manpower requirements. A family of curves has been generated, plotting LIF against man days per $1000 of construction cost for each of three labor rates. Use of these curves requires the determination of a suitable LIF. The average LIF for a project may be determined from the cost-summary chart. For specific tasks of large size, such as feeder work or panel work, the LIF applicable to the particular task should

MANPOWER REQUIREMENTS

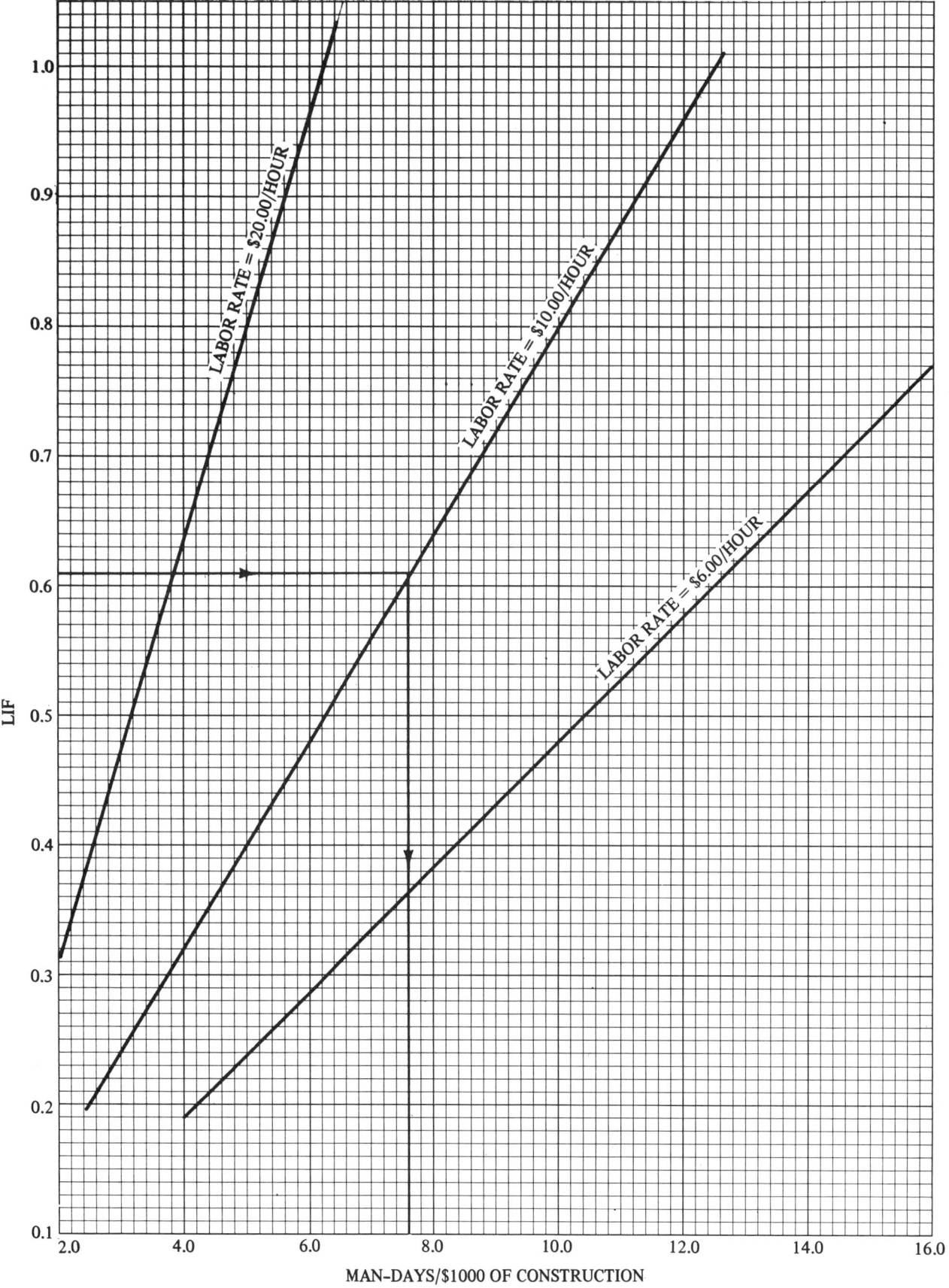

Fig. 3-12.

be used. To find manpower requirements, enter the chart with the LIF, proceed to the proper labor rate, and then read the man-days per $1,000 of construction required (see arrows on the chart for an example).

Information on labor requirements will allow the electrical designer to assist in the development of a master time schedule for the entire project. Isolation of tasks with certain priorities or on the critical path will allow those tasks to be assigned a time span based on information assembled in the manner previously described. For example, part of the CPM diagram may look like this:

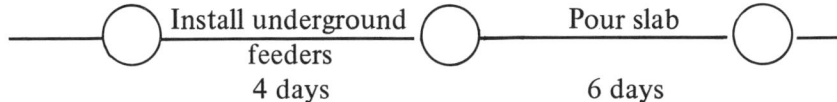

In order to provide a crew of sufficient size to complete the task in the required time, the designer should ascertain the total time required from information in his estimate. Crew size is the result of dividing man-days required to do the work by the time available in which to do it.

When scheduling demands increased manpower loading or overtime work, efficiency will suffer. Nominal overtime has little if any effect on productivity, but a regular schedule of overtime will cause an observable drop in efficiency. Where these losses cannot be avoided by schedule adjustments, factors should be considered for their effects on the effective labor rate. The results will be in accordance with factors in Fig. 3-13. In this case overtime labor rates have to be used, since these invariably are already at a higher rate than straight time.

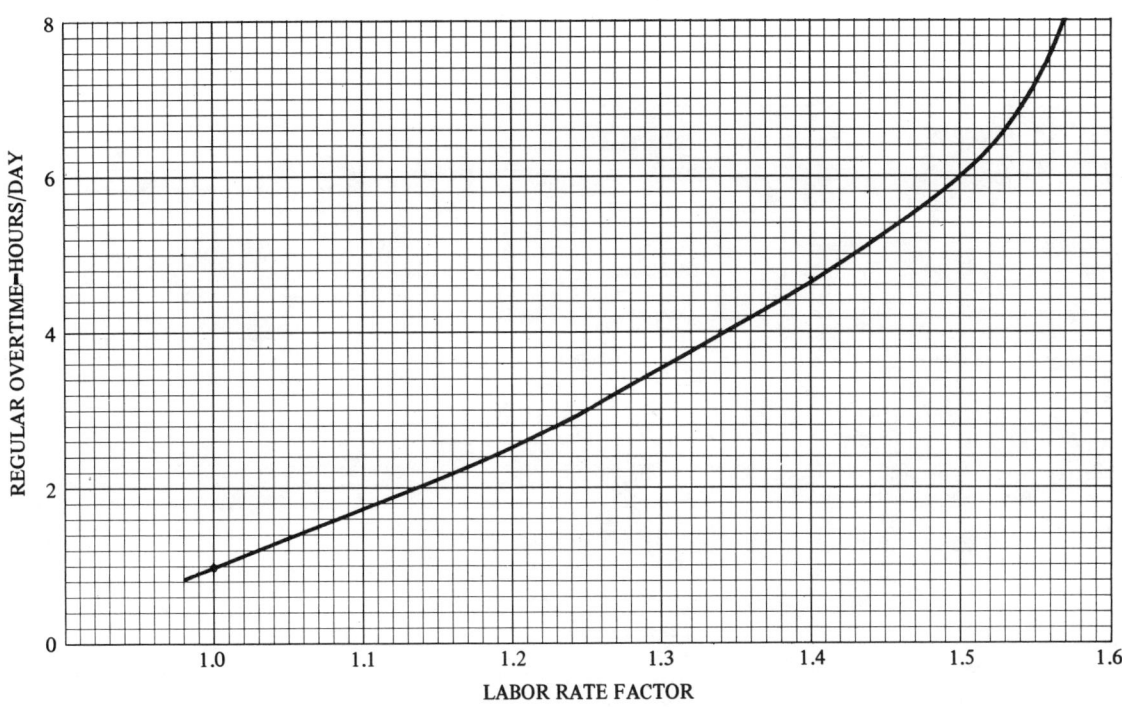

Fig. 3-13.

BOARD OF EDUCATION. CITY OF NEW YORK
BUREAU OF CONSTRUCTION

SCHEDULE OF ITEMS AND COSTS

School _____ Borough _____

Contractor _____ Architect _____

The total cost of each of the following items and sub-divisions thereof, must be inserted in the columns so captioned opposite same. "Costs" shall represent the true value of the work installed and in addition contain in each item the pro rate share of the contractor's overhead and profit. The space opposite items not included in the contract are to be left blank. Any items of work included in the contract, but not listed, must be inserted in the numbered blank spaces provided.

This Schedule must be dated and signed by the contractor or an officer of the firm in the space provided and returned to the Director (Division of Design and Construction) as provided for in the contract.

ITEM NO.	ITEM	COSTS NEW BLDG. OR ADDITION	COSTS EXIST. BLDG. MODERN	TOTAL COSTS	%
1A	Insurance				
B	Bond Premium				
2	Temporary Light and Power Installed				
3A	Conduits—Basement Floor				
B	First Floor Slab				
C	Second Floor Slab				
D	Third Floor Slab				
E	Fourth Floor Slab				
F	Roof Slab				
4A	Nippling—Basement				
B	First Floor				
C	Second Floor				
D	Third Floor				
E	Fourth Floor				
F	Roof Conduit				
5A	Conduits between Main Switchboard and Panel Box Locations including back boxes				
6	Conduits between interconnecting Box Locations				
7	Conduits between Sound Control Cabinet and L.S. Box Locations				
8	Conduits to TV Rack, between TV outlets and TV interconnecting box				
9	Conduits between Clocks and Terminal Boxes				
10	Conduits and Boxes in Hung Ceilings				
11A	Conduits—Underground Electric Service				
B	Conductors Service Feeders				
12A	Fire Alarm System—Exterior Conduits and New Manholes Installed				
B	Exterior Cable installed				
13	Telephone Conduits Installed, Interior System				
14A	Wiring for Light and Power—Main Feeders				
B	Branch Circuit Wiring				
15	Wiring for Sound				
16	Wiring for Clock System				
17	Wiring for Fire Signal System—Interior, and Smoke or Heat Detector System				
18	Switches, Receptacles and Plates				
19	Fire Signal and Detection System Equipment, and Control Boards				
20	Interior Telephone Switchboard, Equipment, and Back Boxes				
21A	Panel Boards—Light				
B	Power—Shops, Fan Rooms, Boiler Room, etc.				
22	Stage Switchboards and Dimmer Board Installed Complete				
23	Motor Starters and Controls, Motor Control Centers				
24	Main Switchboard, Sub-Distribution Center installed and Wired Complete				
25	Electric Service and Metering Equipment				
26	Loud Speakers and Clocks				
27	Sound Control Cabinet Installed and Wired Complete				
28	Television System Rack, TV Outlets, Installed and Wired Complete				
29	Program System Equipment Installed, Wired and Tested Complete				
30	Time Recorder. Complete (Total number. . . . . . .)				
31	Emergency Lighting System Equipment Installed and Wired Complete				
32A	Light Fixtures—Basement				
B	First Floor				
C	Second Floor				
D	Third Floor				
E	Fourth Floor				
F	Cafeteria				
G	Auditorium				
H	Gymnasium				
33	Lamps and Glassware				
34A	Stage Lighting—Border Lights				
B	Soot Lights				
C	Proscenium Lights				
35	Language Lab Equipment Installed and Wired Complete (. . . . . .Rms)				
36	Sprinkler Alarm System, Complete				
37	Patching and Painting				
38	Misc. Electrical Equipment—Kilns, Grinders, Eraser Cleaners, Dental Equipment				
39	Removals				
40	Miscellaneous Items				
41	Testing, Adjustments and Certificates				
	TOTALS				

Contractor _____ Date _____
(Signature of Contractor or Officer of Firm)

Architect or Engineer _____ Date _____

B. of E. Chief of Electrical Design Section _____ Date _____

Fig. 3-14.

PAYMENT SCHEDULES

A not insignificant task required of the designer-estimator is the establishment and/or approval of payment schedules or requisition schedules. Most contractors are in construction for the money. They tend to want their money as early as possible in the project. Several schemes exist to regulate cash flow. These range from the sophisticated CMCS programs in use by GSA (matching payments automatically to definable work produced) to payment on requisition. Any system requires cost evaluation over the life of the project. This will ordinarily break down into early and late types of tasks. Obviously, conduit work will precede lighting fixture work, and so on. The designer-estimator's job is then to assign these items fairly in time and to determine the anticipated cash flow from his estimate. The charts in Chapter 5 should help formulate cash requirements in the very early stages of planning.

Payment schedules may be based on a list of items similar to that of the cost model. If convenient, these may be expanded into more and more detailed itemization in the later and final estimates. If found desirable or if required, these items of cost may be related to work items in the detailed or final CPM schedule. An example of a detailed schedule is given in Fig. 3-14, which is a cost schedule for electrical contract work prepared by the New York City Board of Education. A schedule such as this would be filed by the contractor shortly after the contract is awarded. After verification by the design engineer, the schedule is resubmitted monthly by the contractor with his request for payment. The request would list work completed in the billing month with its value as a percentage of the total value of each line item.

4.
Building Components

SUMMARY

In order to use the pricing charts, it must be clear exactly what items of work are included in each chart. For our purposes we have begun this division by showing a building as consisting of two major components. These are the *basic building* and *other systems*. The basic building is represented by the entire electrical distribution system. Elements of the basic building are shown on a simplified one-line diagram. These parts are identified so as to key into descriptive information in this chapter. Each item is defined, the scope of work is laid out, and its contribution to the electric power load is discussed.

Another use assigned to the diagram here is as a guide in assembling electrical load information. Accurate estimates of electrical service can be made only if the electrical load requirements are known. The charts are divided into two types, *load* and *cost*. The load charts can be used to determine total load within the framework of the diagram.

Certain special areas are also discussed.

BASIC BUILDING—ELECTRICAL

For our purpose, we define the basic building as all electrical work required to bring power from the building line to the power-consuming devices. Expanding on this, we have electrical service, metering, distribution, feeders, panels, and outlet boxes. Without these no building can function. But, to be useful the building will have additional electrical equipment: lighting, motors, and communication and possibly security systems. In addition to these things, rural buildings and even some urban buildings might have extensive site work and lightning protection. For the picture of the work to be complete we have to introduce operating costs of the electrical contractor, overhead and profit, temporary light and power, removals if necessary and costs of maintaining the job. All of these items will be organized and dealt with.

The electrical one-line diagram of a building can be thought of as a tree. It is anchored to the utility company supply lines, rises in a main trunk or series of main trunks, and then branches with more numerous divisions. As in a tree the size of the main trunk and root is determined by the burden of the branches. We have devised simple methods for accumulating the electrical load of the branches and tak-

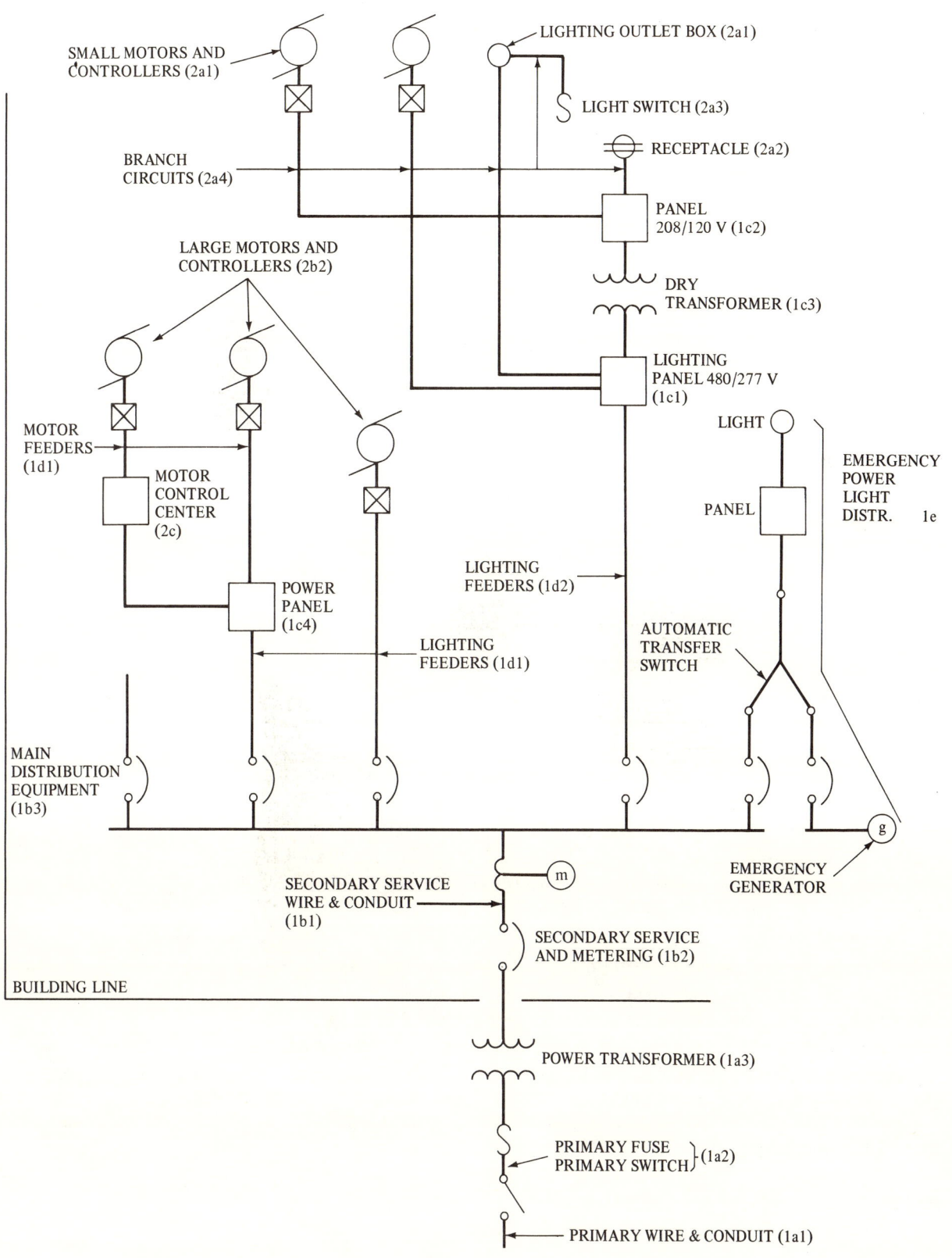

Fig. 4-1.

BUILDING COMPONENTS 89

ing them together to determine the size of the main electrical service. All this information will come from the charts or tables.

Three basic electrical characteristics are voltage, amperage, and power. Voltage is usually determined by what the utility company delivers and can be thought of as a potential force. Voltage is present and available at a receptacle but does no work until some device is plugged in. This is equivalent to head in piping systems. Amperage is a rate of current flow, and can be thought of as similar to gallons per minute flowing in a pipe. Power, measured in watts or kilowatts (KW), is proportional to the product of voltage and amperage, and thus would be equivalent to the head times the rate of flow.

The system of charts generally uses watts per square foot as one of the coordinates. This permits the user to assemble data on the power-consuming devices to determine the ratings of other equipment. Once the equipment has been sized and recorded the cost charts are used to obtain relevant prices. This method relates the equipment costs directly to equipment size, which in fact is the true relationship. Costs given on a square-foot basis without reference to electrical load per square foot cannot be accurate in pricing the individual pieces of equipment.

First we will lay out a bare-bones electrical diagram of a basic building (Fig. 4-1). We will follow the diagram outline in assembling the estimate, since this will put the whole thing in a logical sequence.

Cost per Square Foot

All costs used in the charts or tabulations are in dollars. Areas considered are gross areas. Equipment spaces and other special spaces considered as less than typical in area weight factor are considered in accordance with definitions illustrated in Fig. 4-2. As can be seen, all square feet are not equal. In using charts, some caution should be exercised in the consideration of the area multiplier to use. Space of mixed use or occupancy should be gauged accordingly. The accuracy of an estimate will be greater if smaller spaces are used and more specific unit prices are developed. Where the state of the drawings and specifications permit more detailed considerations, these should be followed.

The following descriptive material defines items included in each of the lines on the reporting form. These items correspond in their identifying number-and-letter key to the items illustrated in the basic electrical building diagram (Fig. 4-1). To keep with our analogy of the tree, we will start collecting the loads at the branches and carry them to the utility company service.

2a1 Lighting Outlet Box

Definition

This is the point of connection of the lighting fixture. Every lighting fixture is wired to a box whether the fixture is recessed or surface-

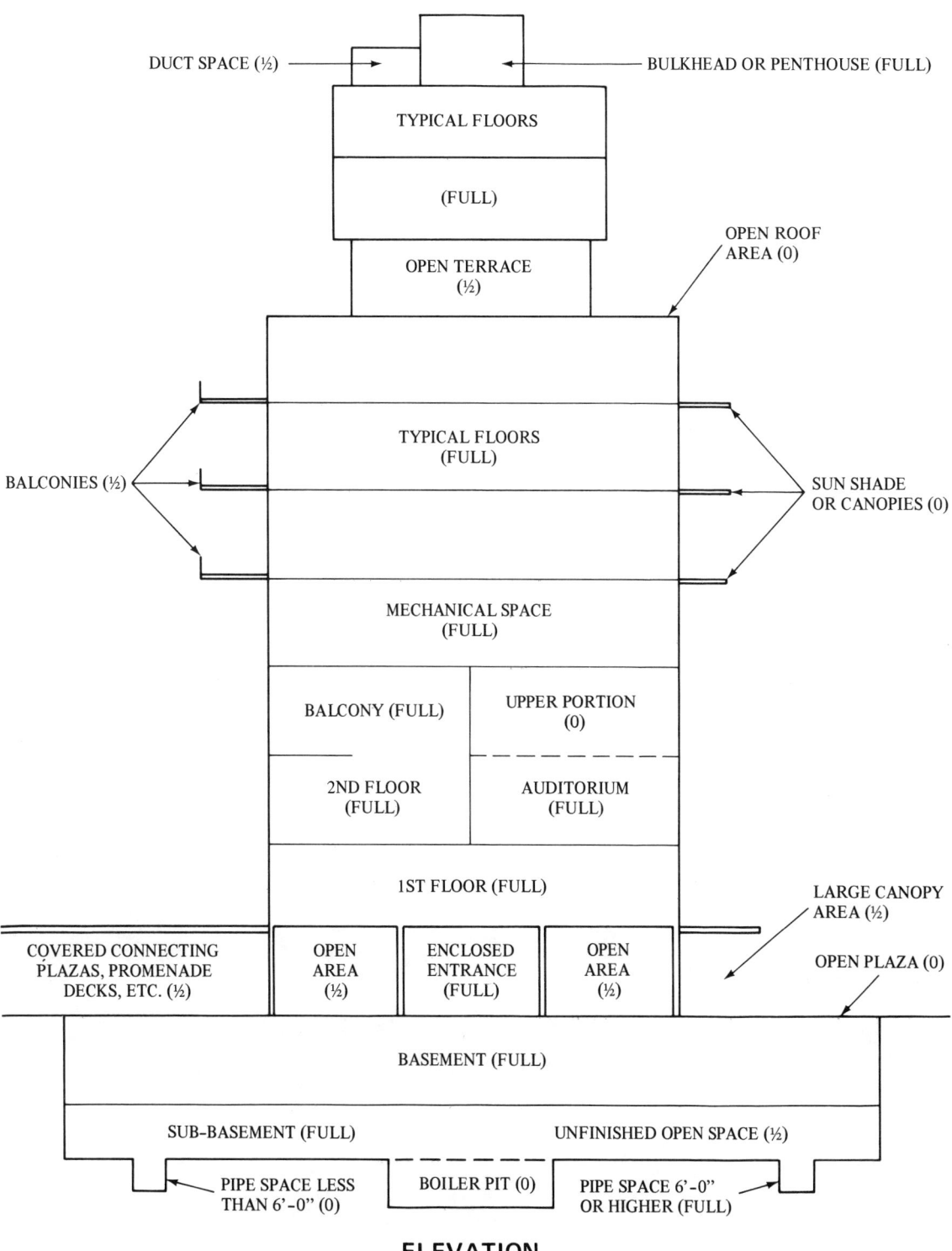

FACTORS FOR CALCULATION SQUARE FOOT AREA

ELEVATION

Fig. 4-2.

BUILDING COMPONENTS 91

mounted, fluorescent, mercury, or incandescent. The box serves as a convenience in initial installation, and as a point from which future changes may begin. Boxes generally in commercial and institutional work are made of pressed steel. In jobs with hung ceilings, the wire connection between the box and the fixture is run in flexible metallic raceways. Costs of this item depend directly on the lighting level and the type of fixture used. Higher lighting levels using any fixture type require more lighting fixtures and consequently more boxes. So too, fixtures of lower efficiency will require more fixtures for a required lighting level and again more boxes. These characteristics can be seen in the Chapter 5.

Electrical Load

Electrical load represented by lighting depends on the fixture type and lighting intensity. The choice of lighting level and fixture type is one of aesthetics as well as one of visual requirements. Relative efficiencies of fixtures can be seen in the charts. Life-cycle effectiveness is something far more subtle than may be presented in a simple chart and should be analyzed separately when the time comes. Insofar as choice of lighting level for seeing purposes is concerned there is much controversy. Remember that with any fixture or lamp type more footcandles will require more fixtures and lamps. This of course requires that more energy be provided to burn the lamps. The fallout of this is the requirement for extra air conditioning to remove the extra lamp heat. Lighting will, in most cases, represent the single most expensive electrical item in the life of a project. The cost of operating lighting systems will often exceed the costs of operating heating and cooling systems. The choice therefore should be made carefully and with due consideration to functions served by the lighting system. However, once the selection is made the charts will indicate the electrical load in watts per square foot to be carried on the reporting form.

Work Included

Costs for this item include only the materials and installation of the box itself. The cost of fixtures will appear as a separate item and is not a basic building cost.

2a2 Receptacle

Definition

Included in this item is any kind of fixed wiring device used for the connection of equipment with flexible plug-in electrical cords. An example is the ubiquitous 120-volt grounding-type receptacle found in every room in a house in increasing numbers with each code revision. Then there would be more exotic types such as those used for portable x-ray equipment, electric ranges, heating equipment, and so on. Our charts are based on the average of all receptacle types in a particular building type. These devices are always installed in boxes, which are usually made of pressed steel. Outdoor devices are usually found installed in cast-aluminum boxes. Insofar as quality of the device is concerned, always use the specification grade or the mil-spec equivalent. Devices are such a small part of the cost structure that in our view it is foolish to use a cheap device. Almost all receptacles, and absolutely all of the 120-volt domestic type, are equipped with a third pole for grounding. Where a metallic raceway is used with metal boxes, the raceway itself is usually an acceptable grounding system and a grounded-type receptacle is used. This is merely a receptacle that has a metal strip along the back connecting the grounding pole to the box by means of the mounting screws. In a nonmetallic conduit system, grounding-type receptacles are used. These allow grounding by the connection of a ground wire, run with the circuit wires, to a grounding screw on the device.

Electrical Load

Hospitals and residences have the strictest code requirements for number of receptacles. Other building types generally are laid out for convenience. Although receptacle circuits may usually be lightly loaded, there are times when the reverse is true. For this reason the watts-per-square-foot chart considers receptacle circuits loaded at 80% of their nominal rating. This figure is used in determining panel sizing and branch-circuit conduit requirements. The loads are for normal utility purposes and do not include electrical heating more than the occasional bathroom heater.

Work Included

This item includes materials and labor for the device and the box in which it is installed, along with a complementary finish plate.

2a3 Light Switch

Definition

This item includes all local toggle switches controlling lighting. Switches of 120 volts and 277 volts (the two usual voltages used for lighting) are included. Generally the 277-volt system will be used in larger buildings with 480-volt three-phase utility-company service. Included also will be the so-called three-way switch used for controlling lights in a room from two locations. Low-voltage switching using 24-volt relays to control lighting is a special item and will not be part of this cost item. Switches are always installed in boxes, which are usually of pressed steel. Outdoor devices are usually found in cast-aluminum boxes. Use a good grade of switch: specification grade or equivalent.

Electrical Load

Switches are control devices not consuming devices and so do not contribute to the electrical load.

Work Included

Included in this cost are materials and labor to install the box, device, and finishing plate.

2a4 Branch Circuits

Definition

Wire and raceways from the lighting and appliance panels are called *branch circuits.* They feed electricity to all lighting fixtures, wiring devices, and small motors. Wire sizes generally are #10 and #12. Insulation is usually a thermoplastic type such as THW. Wires are made of copper. Aluminum wire in these sizes has pretty much been written out of the National Electric Code because of fires due to improper installation. Wires may be in metal-armored cables (BX), in thermoplastic protection such as Romex, or pulled through raceways. Raceways consist of rigid steel galvanized conduit, intermediate metallic conduit, electrical metallic tubing (EMT), flexible wound steel such as Greenfield, or plastic tubing among the more common types. These have been listed in order of descending cost. Rigid and intermediate grades are most suited for use in poured concrete, EMT for exposed, non-hazardous areas. Plastic, BX, Romex, and Greenfield may be used in concealed locations such as partitions and hung ceilings. There are certain code restrictions on the use of these materials.

By and large the density of branch circuit work depends on the lighting level. This determines part of the total, the other part being determined by the receptacle requirements. Putting lighting and receptacle requirements together gives us figures for branch circuit work per square foot of space. Our figures consider the 480/277-volt and 208/120-volt lighting systems usually employed.

Electrical Load

Branch-circuit work itself is non-power consuming and so does not contribute to the electrical load.

Work Included

This item takes care of the labor and materials needed to get electricity from the panels to the boxes in which consuming devices are connected. Included are wire and conduit and their installations and conduit connections at both ends of each run. For convenience wiring of small motors is included in this item. Not included is labor for wire connections at the ends of each run.

2b Motor and Controller

Definition

In this cost item we consider all non-process motors between very small and very large. (Process motors such as those used in factories should be priced on an individual basis.) In this category we have motors usually up to 75 horsepower. Generally these are the motors fed from power panels. Large motors often have their own heavy feeders from the main distribution equipment, and smaller motors (up to about 7½ horsepower) may be conveniently connected to lighting and appliance panels.

Generally, non-process equipment is served by motors operating 24 hours per day or for the daily period of building occupancy. Middle-sized motors used for this type of duty require auxiliary devices called *starters*, usually magnetic across-the-line starters. These are one-step, one-speed, control devices that will permit remote electrical or automatic starting of the motor. They provide motor overload protection but no short-circuit protection. Larger motors usually cannot be started in one step and require starters that are both more complex and more expensive.

Electrical codes require that disconnecting devices be located within sight of the motor and starter. These are mainly intended as a way of protecting personnel from electrical shock or injury while working on motors or starters. In addition to overloads, motors must be protected against short circuits. This is usually provided by the disconnect as a second function. Protection is in the form of fuses or circuit breakers. A so-called combination starter will include the disconnecting means, short-circuit protection, overload protection, and starting functions in a single enclosure.

Electrical Load

In the charts we have listed the two categories of motor load that can be predicted with some accuracy. These are motors associated with air conditioning and heating equipment, and general non-process motors such as water pumps, elevators, etc. Process motors vary greatly from industry to industry. The extent of process horsepower should be obtained from the client. Then this load can be added along with other loads to determine the size of the electrical service and its cost. Once motor horsepower is obtained, average motor feeder prices can be found in the tables.

Work Included

Customarily, starters are furnished by the contractor supplying the motor. Disconnects are furnished by the electrician. Aside from the setting of the motor, this other work is usually in the electrical contract.

2c Motor Control Center

Definition

This is a grouping in a single structure of motor controls similar to those described in 2b. Equipment requiring interconnection may be connected at the factory. Other controls and pilot lights may be added. Motor control centers become economical where motors are located in the same general area and perform functions requiring that they be interlocked. Motor control centers also serve as a maintenance convenience. Physically, the individual motor controls will be assembled in a structure containing bus bars, wiring space, and structural supports. The units may be shipped in stacks approximately 80 inches high by 18 inches wide, which are assembled on the job. Each unit serves as a distribution point to feed power to the motors in its group. A single larger feeder will bring power to the motor control center.

Electrical Load

The motor control center itself merely controls and distributes power. It is not a power-consuming device and therefore it has no contribution to the watts-per-square-foot load. Of course the motors served have to be included as part of the total power load. These should be added as outlined under 2b.

Work Included

In contrast to practice for individual motor controls, motor control centers are usually part of the electrical contract. Great care should be exercised in coordinating this work among members of the design team. This is a source of grief on many jobs.

To recapitulate: included in this line is the motor control center, with starters and circuit protection, motor connection, and labor associated with all of these items.

1a High-Voltage Work

Definition

In general building construction usage, high voltage is customarily voltage above 480. Larger projects of the campus type may buy power from the utility at some high voltage, say 13,800, reduce it down to 4,160 volts, and distribute this to the several buildings. Large single buildings may buy power at one of these higher voltages. High-voltage work may consist of any one of several packages, depending on the utility company and the voltage convenient for the job. The switchgean and related equipment can be elaborate, expensive, and varied. Close estimating of this work only becomes possible when it has been detailed, and then usually in conference with an equipment manufacturer. However, we can place an average figure on the equipment using it over the watts per square foot base with a building total load. Along with the other cost factors its effect on the total price is somewhat diluted to help soften bad errors. Our figures are for the average high-voltage work to a single building. For a campus-type project an additional cost of bringing high voltage to each building (which the utility company would do in a single-building situation) must be added as a subsidiary item. This will range around 25% of the cost of individual building switch gear.

Electrical Load

Switch gear and service equipment add no electrical load of themselves (aside from minor heating losses).

Work Included

Our prices use an average high-voltage switch-and-transformer arrangement. Costs in this category are broken down into:

1a1 Primary wire and conduit. This is high-voltage wire, conduit, and trenching beyond the building line. A proper cost can only be obtained when the length of run has been estimated. Unit prices can be obtained from the tables, but of necessity there is no cost-per-square-foot chart.

1a2 Primary switch and fuse. This includes the relevant items serving the building under consideration, without subfeeds to other buildings.

1a3 Power transformer. This is the transformer serving only the building in question, without sharing its output with any other building.

1b Main Distribution Equipment, Secondary Service, and Metering

Definition

Essentially, this would be equivalent to a low-voltage service if low-voltage supply were bought from the utility company. Whether this is the case or the work represents a downstream part of the high-voltage work, the line item considered here is the same. This work commences at the utility service or power transformer and consists of work through the electrical distribution equipment. Connecting the source to the service switch will be the secondary-service feeder (1b1).

There will be some type of main-service disconnect and protection in the form of switch, fuse, or circuit breakers (1b2). Units may be single, or in the case of larger buildings, may consist of disconnect devices of the same type in parallel. This would be preceded or followed by a current transformer to connect to the utility company meters. Metering may also be used, in campus-type projects in each building.

Next in line is the main distribution equipment (1b3). This consists of a fixed assembly of switch fuse or circuit breakers with bus bars and usually a dead-front metal cover. It is at this point that the bulk power available from the utility company lines is broken into smaller sizes. These sizes depend on the load or series of loads they serve. Each load is served power by a feeder consisting of conductors and raceways. Feeders are brought to the main distribution equipment, where each is protected by its own switch fuse or circuit breaker. In addition to protection, these devices provide a means of disconnecting a single feeder from the main electrical power source.

One or more of these feeders may be used to service the emergency electrical requirements in the building. A feeder may be connected to several so-called emergency panels, where emergency lighting and other equipment may be connected. Larger installations make use of generators for this purpose. To prevent the generator from operating when normal utility power is available, both power sources are connected to a selector switch. The common side of the switch serves as the power source to the emergency loads. Normal power flows through a sensing element that keeps the selector on *normal* as long as this is available. On normal-power failure, a signal is sent to start the generator. After this has come up to speed, another sensor tells the selector switch to transfer over to generator power. In this way, power at the emergency panels may be interrupted for 6–10 seconds. Upon the return of normal power availability, the transfer sequence is reversed and the generator is stopped. So-called uninterruptible power supply (UPS) is not considered here but, of course, could be added as a line item if its cost is determined from other sources.

Switch gear and main distribution equipment can become very complex. Complexity may be necessary where extraordinary reliability is necessary, but there is sometimes a certain amount of overkill due to a designer's fancy. Service equipment should not be more reliable than the utility serving the project. A critical eye for unnecessary expense should be used here. Prices in the chart are commensurate with the size of building served. Large jumps noticed, not in dollars but in percentage, should be viewed critically.

Electrical Load

Switch gear and service equipment add no load of themselves.

Work Included

This line item takes the cost of material and labor from 5 feet outside the building through current transformers and metering, main service switch and protection, through the main distribution equipment. On 480-volt service, ground-fault protection is included in the price.

Costs in this category are broken down into:

1b1 Secondary wire and conduit. This is the run from the power transformer or utility company service to the main distribution equipment.

1b2 Secondary service and metering. When purchased from the utility company, this includes work required to establish electric service between the building line and the main distribution equipment.

1b3 Main distribuion equipment, This includes work connected with the main distribution switchboard, not including branch-feeder connections.

1c Light and Power Panels

Definition

These are boxes in which further breakdown of bulk power is accomplished for final utilization. Within the box may be a main disconnecting and protecting device and then a series of small circuit breakers. These now protect so-called branch circuits emanating from this point. Lighting panels are equipped with 277-volt or 120 volt branch circuits, depending on whether the service is 480 volts or 208 volts, respectively. Panels with 120/208 volts may be used to serve receptacle loads and/or motor loads; 277/480 volt panels may be used to serve lighting loads and/or motor loads.

Electrical Load

Panels serve as protective devices and contribute no electrical load themselves.

Work Included

Furnishing and installing the panel, all branch circuit breakers, and the connection of all branch circuits are costs carried on this line.

1d1 Motor Feeders; 1d2 Lighting Feeders

Definition

Feeders basically transport power in bulk. Consisting of wires protected by conduits, they are the long-distance distribution component in the overall distribution system. Power is broken down at the main switchboard into convenient blocks. In order for the blocks of power to be delivered to large motors or to further distribution points, feeders are used usually rated at 208 volts or 480 volts in the building. Where single-phase loads are to be served by the feeder, a fourth wire, the neutral, is carried with the three-phase conductors. In the case of three-phase loads, only three-conductor feeders are required.

Feeders consist of conductors and enclosures. They are wire-and-conduit or sometimes bus structures. Conductors within the building are usually rated at 600 volts. They are made of copper or aluminum. Enclosures and conduits are made of several weights of steel or aluminum. Busways also are made of copper or aluminum.

Electrical Load

Feeders add no electrical load of themselves.

Work Included

Included in this line are all items of labor and material for ordinary feeder runs.

1e Emergency Power and Light Distribution

Definition

Emergency requirements vary from building to building. In this line item all the components required to make a complete system are included. Emergency systems essentially consist of an energy source alternative to that normally available as well as means of detecting failure of normal service and of transferring a part of the load to the emergency service.

The type of load placed on the emergency service will vary with the building type and particular requirements of the user. Hospitals have several orders of emergency systems. Each of these has to be considered in determining the size of the emergency service. In some cases, redundant distribution systems may be required, with various types of contingent backup or sequential starting of several pieces of equipment using either manual or automatic synchronization.

The consideration given here is to systems of relative simplicity. These consist of the generator, its protection, and the automatic transfer of electrical load.

There are two major divisions of emergency power equipment: central and local systems.

Local System. This usually consists of a self-contained battery, a charger, a transfer relay, and a light fixture either attached to or installed near the battery unit. This is a kind of go-anywhere device. It can be plugged into any normally energized, 120-volt, AC circuit. Where wattage requirements are low and emergency lighting locations are widely separated, a system of this type is ideal.

Another type of local system is the so-called transistorized-ballast type. This device actually contains an inverter changing its DC battery voltage to AC so it can operate a fluorescent lamp. It can be arranged to operate one or two lamps in a fixture already forming part of the ceiling pattern. Its advantage is that no additional fixtures are needed in those areas where extra fixtures may be aesthetically undesirable. Where this feature is found important, be prepared to pay somewhat more than for the straight DC battery pack and light.

Central System. Here too there are two major divisions into types. There are the DC central system using 32 volts or 110 volts DC and the central AC-generator type. In large installations, those requiring extensive lighting and some emergency power, there is really no choice other than the AC generator. In this case economic choice is limited to competition between suppliers. Try not to exclude a manufacturer by inadvertently specifying a feature unique to one product only. These are large-cost items, and competition between suppliers should be encouraged.

In considering a choice between generator systems it is important to consider the total package. This price must be sought after, it is not always clearly represented by equipment suppliers. Compare items and quality for all of these components:

1. Fuel storage and handling
2. Flue requirements
3. Engine cooling
4; Engine-room ventilation
5. Auxiliaries, batteries, chargers, alarms
6. Sound isolation
7. Vibration damping
8. Electrical tranfer equipment
9. Equipment ratings (both prime-mover and generator) and load characteristics
10. Spare-parts inventory required

Items not directly related to economic choice but still related to choice should include:

11. Reliability and stability of manufacturer
12. Availability of service organization
13. Demands on in-house service capability

A large central system can sustain the building in which it is used for long periods, operating required motors and lights. Selection of equipment to be connected is generally restricted to required or important loads so as to hold down generator costs.

DC central systems find good application where egress lighting in a building of some size is required. The advantage is that equipment requiring maintenance is centrally located. Vandalism and theft of the alternative individual units is eliminated. These systems will contain a number of 12-volt batteries, battery charger, battery racks, transfer switches, and some type of branch-circuit protection. Because voltages are lower than those usually used for branch circuits, larger wire sizes and reduced wattage are the watchwords. Low-voltage lamps in special fixtures will be required. A separate wiring system is used.

Electrical Load

In computing the building electrical loads these systems essentially add no load. Generators themselves may be selected on a watts-per-square-foot basis using the emergency load requirements.

Work Included

For the local battery pack or transistorized ballast, only the costs of labor and materials for the units apply. These may consist, for example, of a new outlet or receptacle and its installation. In the case of central battery systems, costs include the central battery and load transfer system, emergency fixture cost, and branch circuit cost. We assume the cost of a single outlet to be twice that used for standard outlets.

For the generator system, costs in the chart include cost of drive and generator, controls, auto-transfer switch, generator auxiliaries, and silencer. The costs of distribution, fixtures, outlets, and panels are at the same unit prices as for comparable equipment for normal use. These items should be assembled for inclusion in this line when redundant feeders are used. Not included are ventilation, cooling, flues, and fuel-handling system.

OTHER SYSTEMS

We have defined and examined the basic building, and now proceed to other systems. These will vary more widely in type, function, and cost than items in the basic building. They vary from building type to building type in whether or not they are required at all, and within a single building type there can be great variations in function and price. Taken together with lighting, these costs represent a substantial part of the total building. (Item numbers refer to those used on the reporting forms).

Since the choice of these systems is in many cases open, they are somewhat independent of the total building area. They are more directly related to the area served. Where unit prices apply, they do not necessarily relate to similar parameters. For example, a nurse-call system depends on the number of patient beds; the fire alarm system depends on configuration and occupancy; lightning protection depends on the roof area. Using typical square-foot costs is possible but certainly less meaningful than using square-foot costs for the basic building. In order to estimate costs, simple system block diagrams can be made and the equipment priced from data in the related sections.

3 Illumination

A special category is set up for lighting. Although lighting might be thought of as basic to a building its costs vary so widely that this item has been put in with other systems. The final lighting layout is subject to a host of variable factors: client self-image, community pressure, client information or misinformation, manufacturer's claims (both false and true), energy-conservation requirements, consideration of the visual task, aesthetic considerations, and emotional factors. All of these will affect cost.

In order to prepare estimates of lighting cost the design engineer must solve the first-order problem of illumination level required. This is no small task, but should be attacked and resolved as early as possible. Costs are substantial, and foot-candle level will determine wattage requirements and consequently affect other electrical costs such as feeders, panels, and electrical service. An indirect cost of higher foot-candle levels is the increase in air conditioning requirements and the effect of that increase on adding to the electrical motor load. There has been a trend away from the steady increase in so-called foot-candle requirements. More recent considerations are becoming sensitive to the function served by the lighting system. Energy considerations are becoming more important, with both state and federal governments beginning to impose limitations on lighting designs.

After the establishment of a design criterion the search for an aesthetically pleasing, economical, efficient fixture is begun. The designer should feel obliged to examine fixture test data and make a careful selection, since substantial costs are involved in this item. The cost of fixtures themselves varies greatly, as does the quality. Examination of physical samples as well as consideration of test data should be made.

Electrical Load

The charts presented are based on good-quality fixtures of better than average lumen efficiency. From the charts the load in watts per square foot and costs can be determined for this fixture type. This should suffice for earlier estimates, but should give way to actual quotations and specific test data as the design proceeds.

Work Included

Costs for this item include labor and materials for installing lighting fixtures and he lamps they require.

4a1 Intercommunication Systems—Intercoms

Definition

Intercommunication functions are served by public telephone extension lines, wholly owned private telephone systems (described under telephone systems), or by the pushbutton type of intercom set. The last-named is what we will consider here. These systems are comprised of so-called master and remote stations. The master has the facility of selecting any remote station, while the remote station can only call the master. A great deal of flexible instant communication is available where several masters are connected to a system. These systems have the advantages of no-dial calling by depressing a single station button and of no-hands reply. They find good application in control-desk calling to waiting rooms and in calling personnel not likely to be at a desk. Where outside contact is not required these devices deserve consideration against the two types of telephone systems.

4a1 Intercom System—Desk-Type Master Station

System	Cost
10-station master	$2,000
20-station master	$3,700

LIF = .45

Note: These system prices are based on a 50-foot run between remote stations.

Electrical Load

With solid-state electronics, power consumption is minimal. Units plug into any 110–120-volt AC outlet and consume on the order of 50 watts. No consideration is required in loading on central power equipment.

Work Included

Amplifiers and pushbutton assembly comprise the common type of master station. Units are now solid state and very compact. A 30-station master unit is about 14" × 10" × 6" and conveniently sits on a desk. Remote stations are either wall or desk mounted, and all contain call origination facilities. Systems listed here contain a remote station for each master position, wiring, and conduit.

4a1 Intercommunication Systems—Telephones

Definition

For our purpose, the public telephone system includes all work performed in and for the installation and operation of a public-utility-owned telephone system. For large installations there are many services offered by telephone companies. A great deal of flexibility can be built into modern equipment. Telephone-company representatives appear to be very cooperative in describing and suggesting suitable equipment. This of course may be at least partly due to the pressure of alternative equipment available.

In choosing between wholly owned equipment purchased by the owner or equipment leased from the telephone company several factors should be considered:

1. Life-cycle costs
2. Availability of initial capital
3. Flexibility of operation
4. Maintenance facilities.

Estimates of costs should include conduit and boxes where systems are to be installed in this manner. Where convenient, telephone lines can be run in hung ceilings and partitions without conduit. Costs include conduit, outlet boxes, box covers, and terminal boxes. No materials or labor are included for wires or equipment.

4a1 Telephone System

Cost per Outlet = $80.00
Assume 40 feet EMT run average
LIF = 0.80

Electrical Load

Central equipment will require some electrical energy. This load is too minor to affect electrical service equipment.

Cost per outlet = $80.
Assume 40 feet EMT run average.
LIF = 0.80.

Work Included

Utility-owned equipment is usually installed by utility-company personnel. Where empty conduit is used this is part of the electrical contract work. Telephone instruments will require outlet boxes. Terminal cabinets should be conveniently located to minimize branch-circuit home runs.

4a2 Sound Systems

Definition

This item refers to public address and school sound systems. Auditorium sound can be very specialized and is not included. Public address systems are zoned sound-distribution systems. In the school situation the zone consists of a single classroom. In this case there is also usually a provision for originating a call or talking back from the remote station. Central equipment consists of an amplifier with room selector keys and talk-back speaker. Provision is usually made for several input modes, such as microphone, tape, or tuner. Equipment is solid state and therefore very compact. A 20-station unit rated at 35 watts is about 6" × 16" × 10". The minimum system will require this amplifier switching unit, a desk microphone, remote speakers, call-in plates for intercom operation, wire, and, if required, conduit. Classroom and corridor speakers can be purchased assembled in a baffle box. Speaker power requirements on the amplifier are about 2 watts each. Systems will require at least one microphone and possibly some horn-type exterior speakers.

4a2 Sound Systems—Public Address, Schools

School type	Cost/square foot
Elementary	$0.10
Middle	$0.11
High school	$0.13

LIF = .33

Electrical Load

Electrical loads are light. For example, a 35-watt-rated amplifier with 20 room-selector keys consumes 110 watts at peak power. In considering electrical service and panel sizing this item is disregarded.

Work Included

Costs of these systems include a wire, conduit, and distribution cabinet system. Labor and materials for speakers, amplifiers, microphone, and call-in plates are included. In larger schools, tone generators, record changers, and tuners are included.

4a3 Television Systems

Definition

This item is for off-air television entertainment. Studio-type installations with special-effects generators, monitors, videotape recorders, cameras, etc., are tailor made and can cost any amount the purchaser has available. Because of this, studio work is not included. The usual TV distribution system considered here is for a single building. The system would contain antennas, amplifiers, splitters, conduit, coaxial cable, and terminal devices.

4a3 Television System—Wired Outlets, Based on 200-Outlet System Costs

Building type	equipment	conduit and cable	total	cost per outlet
Hospital	$10,000	$ 8,000	$18,000	$ 90
Nursing home	$ 9,000	$ 6,000	$15,000	$ 75
School	$14,000	$10,000	$24,000	$120

LIF = 0.37

Note: These prices are for average conditions and spaces, equipment installed. Smaller systems require somewhat higher unit costs.

Electrical Load

Power consumption is very small, since most equipment is of the solid-state type. Only amplifiers require any electric input at all, and this is inconsequential in sizing electrical service or distribution equipment.

Work Included

Antennas, amplifiers, splitters, matching devices, EMT conduit, coaxial cable, outlet box and cover, and signal terminal devices are included.

4a4 Clock Systems

Definition

Clocks in a building may be individual units on local circuits, units hard-wired to a central control system, or units locally wired but centrally controlled by a superimposed carrier signal.

Individual clock units represent the least expensive choice. Wiring costs no more than an ordinary duplex receptacle. The clock itself will cost the same as or less than a centrally controlled clock. A problem arises in a building of some size when power to a circuit or to the entire building is interrupted. Clocks then have to be reset manually on an individual basis. In some cases this may be clumsy or very inconvenient. The system is, however, very flexible.

Hard-wired systems are centrally controlled from a master clock. This may or may not be tied into a school program system. Each local clock is wired to a distribution system. Corrective signals are sent from the master clock each hour to keep local clocks in step. In the event of a power failure there is a reserve power feature in the master clock. When power returns, correction is directed from the control unit automatically. This system finds favor in buildings such as schools and hospitals. Alterations require a connection back to an existing clock circuit.

Another central system uses a generator to put a high-frequency carrier wave on the normal 60-cycle house current. Correction to local clocks is effected by this signal. With this arrangement a clock may be plugged into any electrical outlet, even newly installed, and still be in step with the system. Obviously this is the most flexible system. In some installations there may be a problem of electrical interference with other electronic equipment.

4a4 Clock Systems

System type	Cost
Program Master clock	$1,500
Single clock with outlet	$ 85

Add price of conduit and wire system using charts 1d1, 1d2. Average LIF = 0.6.

Electrical Load

Clocks themselves are a very minor load. Central equipment also is light. Generators may require their own circuits. Loads are too small to be considered in evaluating the building total load.

Work Included

Clock, clock outlet, circuit wiring, and central equipment are listed. These items are shown with materials and labor costs.

4a5 Nurse-Call Systems

Definition

These systems are used in hospitals and nursing homes as a method of communication between patient and nurse. Systems break down into two basic types: visual and audiovisual. Visual systems indicate a calling station by means of a corridor dome light and some type of illuminated panel at the nurse's desk. Audiovisual systems add the ability for patient–nurse conversation via hand set or pillow speaker. Some of these systems will include an entertainment feature operating through the pillow speaker. A relay operating button on the pillow speaker permits selection of television channels or radio sound which is heard through the pillow speaker. Various other features are available, such as bed monitoring and emergency call. These are low-voltage systems which can be wired without conduit where convenient.

4a5 Nurse-Call Systems

System type	Cost total	per bed	LIF
Visual: 25 rooms and baths, 50 beds, wire and conduit	$5,000 $2,500 $7,500	$150	0.44
Audiovisual: 25 rooms and baths, 50 beds, wire and conduit	$13,000 $ 3,000 $16,000	$320	0.22

Note: These prices are for average rooms. Price per square foot should be based on the nursing-wing area.

Electrical Load

Usually each separate nurse-call system is on its own branch circuit. This is more for convenience than because of the size of the load. Total contribution from this type of equipment is negligible in sizing the main electrical service equipment or distribution.

Work Included

For the visual-system bedside station, call cord, EMT conduit, wire, dome light, emergency station, and the nurse's desk unit are included. Audiovisual systems will contain the same work plus the addition of TV relay wiring and amplifier units.

4b1 Fire-Alarm Systems

Definition

Equipment for automatic detection, manual alarm sounding, alarm bells and horns, and central equipment rack are the essential elements of fire-alarm systems. Some have added sophistication of alarm recording on hard copy and voice communication. Almost all systems are electrically supervised and have trouble indicators.

Basically the system serves to indicate an interruption of current in the signal-transmitting device, whether automatic or manual. In the larger, coded systems the individual station or related group of stations causes the central equipment to send the coded current to all alarm devices. Where printers and outside fire companies are used the alarm is also sent to them. Source of the alarm—duct detectors, sprinkler flow, smoke detectors, heat detectors, or manual station—is indicated by the audible signal. In certain systems for high-rise buildings, instructions are broadcast to fire areas by means of fire command stations. Some smoke-detecting systems may be tied into ventilating systems. These will stop some fans and operate some exhaust fans. Requirements vary with different building codes.

4b1 Fire-Alarm Systems

System type	Cost per square foot
Sophisticated system: high-rise office buildings, hospitals	$1.00
Moderate system: schools, colleges	$0.50
Simple system: banks, stores	$0.25

Average LIF = 0.5.

Electrical Load

A minor amount of energy is consumed in these systems. Their effect on service equipment and panels is negligible.

Work Included

All labor and materials for the associated item are included in the cost schedules. Conduit and wire between stations can be a variable, but for general considerations a typical unit is indicated. Frequently conflicts develop in coordination of equipment, labor, and central equipment when duct-type smoke detectors and fan controls are used. Specifications are sometimes unclear between areas furnished by HVAC and electrical contractors.

4c1 Lightning Protection

Definition

Essentially, lightning protection serves to minimize the damaging effects of lightning. It does this in two ways: first, it reduces the incidence of strikes, and secondly, if lightning should strike the building, it provides a safe path to ground for the energy. Basically, these systems are simple, consisting of air terminals, down conductors, and a good ground.

The choice of whether or not to install a lightning protection system depends on the type of building and its location. Insurance companies will be of help in determining requirements by their rate structure. Where systems are installed they should conform to Underwriters requirements.

4c1 Lightning Protection System

Item	Materials costs dollars	Materials costs unit	Labor costs man-hours	Labor costs unit
24″ air terminal	40.00	each	1.0	each
Copper cable, 59,500 CM	1.38	lineal foot	33.0	1000 lineal feet
Aluminum cable, 98,500 CM	.74	lineal foot	38.9	1000 lineal feet
Ground rod, 10′, connected	19.00	each	2.5	each

LIF = 0.33.

| | | Lightning Protection System—Estimate | | | |
| | | Materials | | Labor | |
Item	Quantity	unit cost	total	unit hours	total
air terminals					
cable–roof					
cable–down					
ground rod					
Subtotal					
Hardware, 15%					
Totals man-hours × rate =					
Total cost:					

Electrical Load

There is no load contribution from this item.

Work Included

This item includes the cost of all materials and labor required for a typical system, that is, air terminals, mounts, collector conductors, down conductors, and grounding. Note that square-foot costs are based on roof area.

Requirements

Roof configuration is a determining factor in the layout of a lightning-protection system. In principle, high points and edges are the locations for air terminals. Typical spacing is specified by Underwriters. For example, a flat roof using air terminals 24 inches or higher will have them spaced 25 feet apart on the perimeter.

Where a roof dimension exceeds 50 feet, intermediate rows of air terminals should be installed, with rows not more than 50 feet apart and with air terminals at least 50 feet on centers. Bare roof conductors (collector conductors) of copper or aluminum are used to interconnect all the air terminals. Roof conductors are connected to down conductors. At least two down conductors are required, located so as to be as widely separated as possible. Where the building perimeter exceeds 250 feet an additional down conductor is required for each 100 feet of perimeter. Down conductors terminate at the ground in a connection to a ground rod or other grounding equipment. The usual system uses a rod not less than 10 feet long driven vertically into the earth. In dry soil or rock conditions more expensive arrangements will be required.

Other materials required for the system are essentially heavy-duty nuts, bolts, and clamps. A variety of hardware is available to mount air terminals to different parapets.

4c2 Temporary Light and Power

Definition

This is the entire electric power and light system required for use during the construction of the project. Equipment will include service connections, transformers, meters, high-voltage switch gear if necessary, secondary distribution equipment, feeders, panels, branch circuits, light and power outlets, and lamps for the duration of the job. Strings of lights have to be moved frequently as work progresses. Distribution equipment should be set up in a place from which it will not have to be relocated until permanent wiring can be used. Sufficient light has to be provided to make working conditions safe. Requirements are spelled out by OSHA. Each space will require lighting as it is enclosed. Usually, outlets are not required to feed more than a 5-horsepower motor.

Electrical Load

Since the system is removed after use it has no contribution to the permanent load.

Work Included

All labor and materials to furnish, install, maintain, and remove the temporary system are included. An electrician is usually required whenever the system is energized. The costs do not include the cost of energy to operate the system. This cost is usually paid directly by the owner and should be considered when estimating total project costs. Feeders required for special testing prior to the operation of the permanent distribution system are not included in the cost shown. This work should be carefully assigned in the specification, especially where large equipment may have to be tested in the early phases of construction.

4c2 Temporary Light and Power—Typical System Cost

Building type	Cost/square foot
Apartment house	$.10
Hospital	.20
Museum	.16
Nursing home	.13
Office building	.30
Parking garage	.10
Public school	.33
Student union	.33

LIF = 0.40.

SPECIAL AREAS

Assigning an electrical cost to certain special areas is almost like asking the question, "How high is up?" By and large the areas considered here are very expensive on a square-foot basis. Fortunately, they do not usually represent a large part of the total installation containing them. Preliminary estimates can be based on preliminary lists given. Later estimates should be performed on a takeoff basis to ensure accuracy.

Hospital Operating Rooms

Operating rooms normally contain an isolated power supply, x-ray outlets, outlets for the surgical lighting, clocks, a communication system, and possibly provision for television equipment. Much of this equipment is available in single-panel assemblies. All nonconducting metal equipment should be grounded, including the conductive flooring. Electrical codes covering operating rooms and other hospital-type facilities change regularly. Be sure your code book is current. Costs should cover the cost of the suite, including adjacent service areas.

Hospital Cardiac Care Units And Intensive Care Units

Recent observation of the lethal effects of stray microcurrents in the human body have resulted in stricter codes for CCU and ICU areas. Stricter for present purposes here can be translated as more expensive. These special areas require specific observation of grounding techniques. A common, nonelectrical ground is used for all metallic equipment in the room. This is tied to the usual green ground system. Circuits in these areas will be on critical and life-support branch circuits. Receptacle requirements are heavy. Physiological monitoring is often provided, but is not considered here as part of electrical costs. Costs should cover the area designated as the CCU or ICU, and so include both the patient and nurse areas.

Kitchen

Electrical equipment continues to represent an increasing part of modern cooking equipment. Frozen-food preparation using reconstituting ovens, warming equipment, radar ranges, etc., all have served to increase electrical use in kitchens. Heavy loads remain the fryers, ranges, and ovens. Kitchens should have their own electrical panels for equipment therein. Each item of equipment generally is put on a separate circuit. This is convenient from the operation point of view in permitting minimum interruption in the event of equipment failure. In the design stage it is a practical way to minimize design drawing changes caused by the usual redesign by kitchen consultants. Kitchen costs can validly be expressed on a square-foot-of-kitchen basis. The cost is proportional to the load, which in turn is proportional to kitchen size. Kitchen power is heavy, and will appear as part of the service requirements. When considering energy conservation methods some of these loads can be conveniently programmed to minimize peak demands. This is a method of reducing electrical demand charges.

Kitchen Electric Cost

Kitchen type	Cost per square foot of kitchen
Sophisticated	$6.00
Medium	$3.00
Simple	$2.00

Average LIF = 0.7.

5. Charts

SUMMARY

This chapter contains the cost and load charts. Each chart is clearly identified as to type. An index is provided listing the subject and line item of each of these charts. Charts have an arrow indicating entry and exit points. Typical buildings of a number of general classifications are spotted on most of the charts. These should serve the designer in locating a cost range within which he can operate. Information not convenient for chart format will be found in the appendix.

Use of the charts should be fairly simple. However, a typical estimate is run through in Chapter 6.

USE OF THE CHARTS

Areas used are gross square feet.

Costs indicated are labor and materials costs without contractor markup.

Costs do not include an allowance for general conditions.

Charts are labeled according to the number-and-letter code used elsewhere in this book. Each chart refers to a line item in the estimate. The same designation will appear in the one-line diagram, the definition section, and the reporting form.

Building types indicated in the charts represent a point on a particular curve which indicates a typical price for that building type. Other building types may be interpolated or may be found in terms of the charted parameters.

Each chart indicates the proper labor intensity factor (LIF) to be used with data in the chart.

Directed arrows on the charts show entry and exit points for their use.

Certain of the charts are load charts. These are presented as a means of assembling the electrical load from the various requirements. Once determined, the loads form a basis for estimating upstream equipment from the proper chart.

CHART INDEX

Lines	Title	Axes
1a1	Primary Wire and Conduit—Cost	Amps vs. C/LF
1a; 1b	Conversion Chart: Total Kilowatts vs. Amperes—Load	Amps. vs. Total Load KW
1a2	Primary Service Switch—Cost	Building Total MW vs. Total Cost
1a3	Power Transformer—Cost	Building Total KW vs. Total Cost
1b2	Secondary Service and Metering—Cost	W/SF vs. Cost/1000 SF
1b2	Secondary Service and Metering—Cost	Total Load vs. Total Cost
1b3	Main Distribution Panel Board—Cost	Total Load vs. Total Cost
1c1; 1c2	Lighting and Receptacle Panels—Cost	Number of Circuits/1000 SF vs. C/SF
1c1; 1c2	Lighting and Receptacle Panels—Cost	Total Number of Circuits vs. Total Cost
1c3	Dry-Type Transformers—Cost	Receptacle Load W/SF vs. C/SF
1c4	Power Panels—Cost	Motor Load/KW/Panel vs. Total Cost/Panel
1d1	Motor Feeders—Cost	Motor Horsepower vs. C/LF
1d1; 1d2	Copper Feeders, 3-Phase, 3-Wire—Cost	Ampacity and Wire Size vs. Cost /1000 Ampere Feet
1d1; 1d2	Copper Feeders, 3-Phase, 3-Wire—Cost	Ampacity and Wire Size vs. C/LF
1d1; 1d2	Copper Feeders, 3-Phase, 4-Phase—Cost	Ampacity and Wire Size vs. Cost/1000 Ampere Feet
1d1; 1d2	Copper Feeders, 3-Phase, 4-Wire—Cost	Ampacity and Wire Size vs. C/LF
1d1; 1d2	Light and Power Feeders—Cost	Watts/SF vs. C/SF
1d1; 1d2	Aluminum Feeders, 3-Phase, 3-Wire—Cost	Ampacity and Wire Size vs. C/LF
1d1; 1d2	Aluminum Feeders, 3-Phase, 3-Wire—Cost	Ampacity and Wire Size vs. Cost/1000 Ampere Feet
1d1; 1d2	Feeder Bus Duct, 480 Volt, 3-Phase, 3-Wire—Cost	Ampacity vs. Cost/1000 Ampere Feet
1e	Emergency Generators—Cost	Watts/SF vs. C/SF
2a1	Lighting Outlet Box—Cost	Sq. Ft./outlet vs. C/SF
2a2; 2a3	Switch and Receptacle—Cost	Devices/1000 SF vs. Total C/SF
2a4	Branch-Circuit Wiring, Light Switches—Cost	Switch Outlets/1000 SF vs. C/SF
2a4	Receptacle Electrical Load—Load	Receptacle/1000 SF vs. Watts/SF
2a4	Branch-Circuit Wiring, Receptacles—Cost	Receptacle/1000 SF vs. C/SF
2a4	Branch-Circuit Wiring, Lighting—Cost	Watts/SF vs. C/SF
2a4	Number of Branch Circuits, 208/120 Volt—Load	Watts/SF vs. Number of Circuits/1000 SF
2a4	Number of Branch Circuits, 480/277 Volt—Load	Watts/SF vs. Number of Circuits/1000 SF
(2b1	Small Motor and Control-No chart required; use same cost chart as devices)	
2b2	Individual Motor Connections (Without MCC)—Cost	Motor Horsepower vs. Total Cost
2b2	Air-Conditioning Motor—Load	SF/ Ton A/C vs. Watts/SF
2b2	A/C Motors and Control (Steam Central Plant)—Cost	A/C Watts/SF vs. C/SF
2b2	A/C Motors and Control (Electric Cooling Plant)—Cost	A/C Watts/SF vs. C/SF
2c	Individual Motor Connections (with MCC)—Cost	Motor Horsepower vs. Total Cost
3.0	Light Fixture—Cost	Watts/SF vs. C/SF
3.0	Light Fixture—Load	Watts/SF vs. Foot-Candles

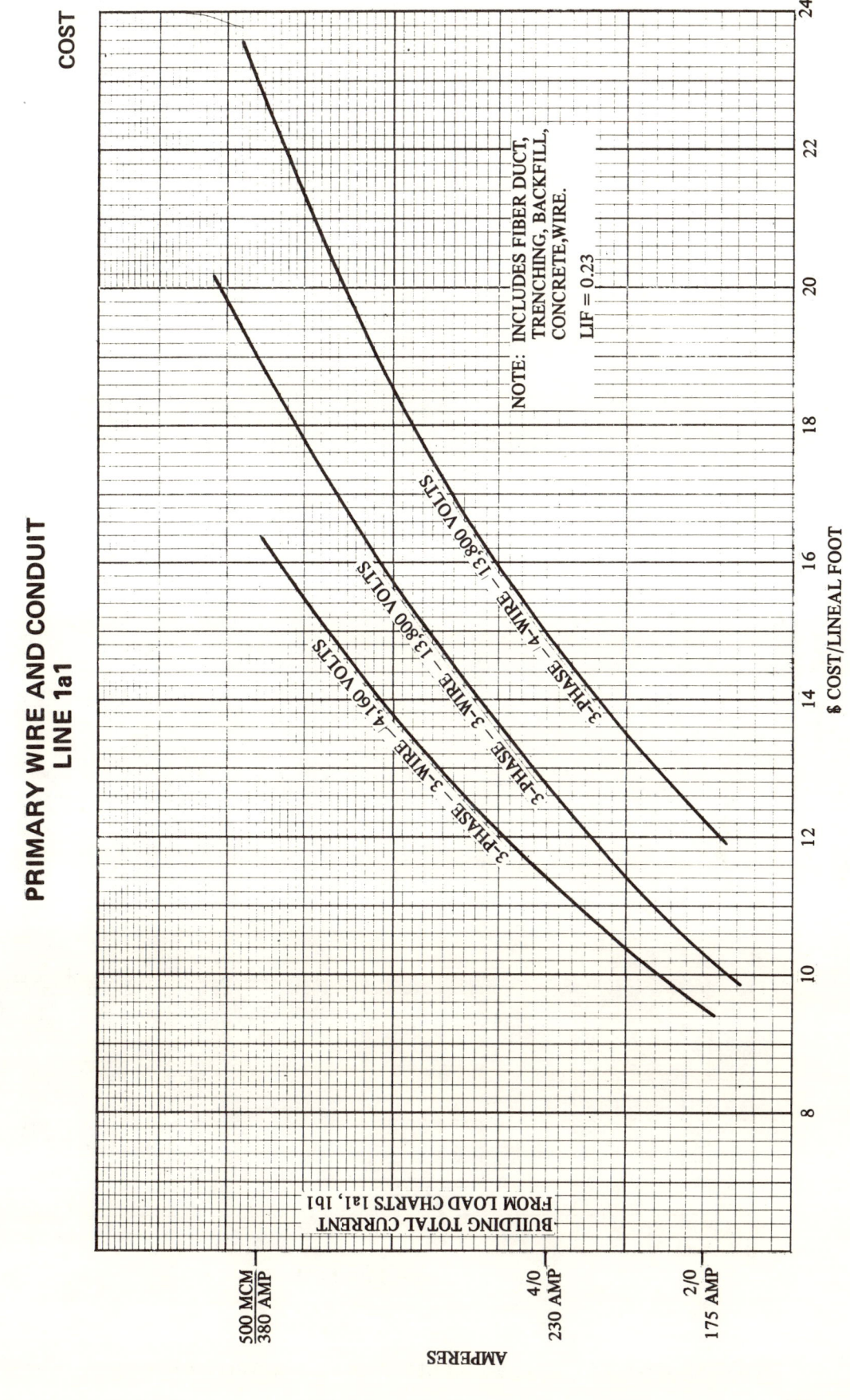

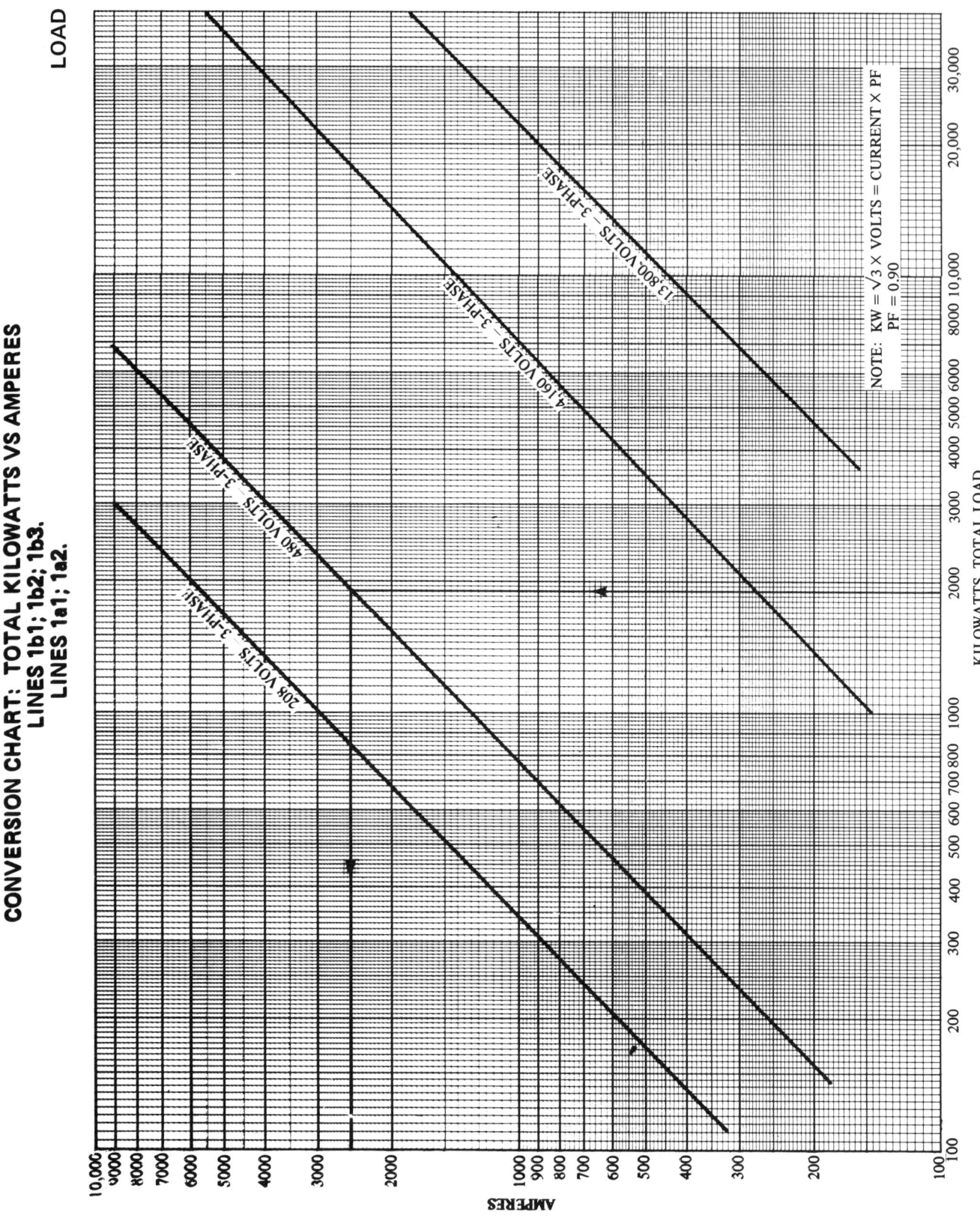

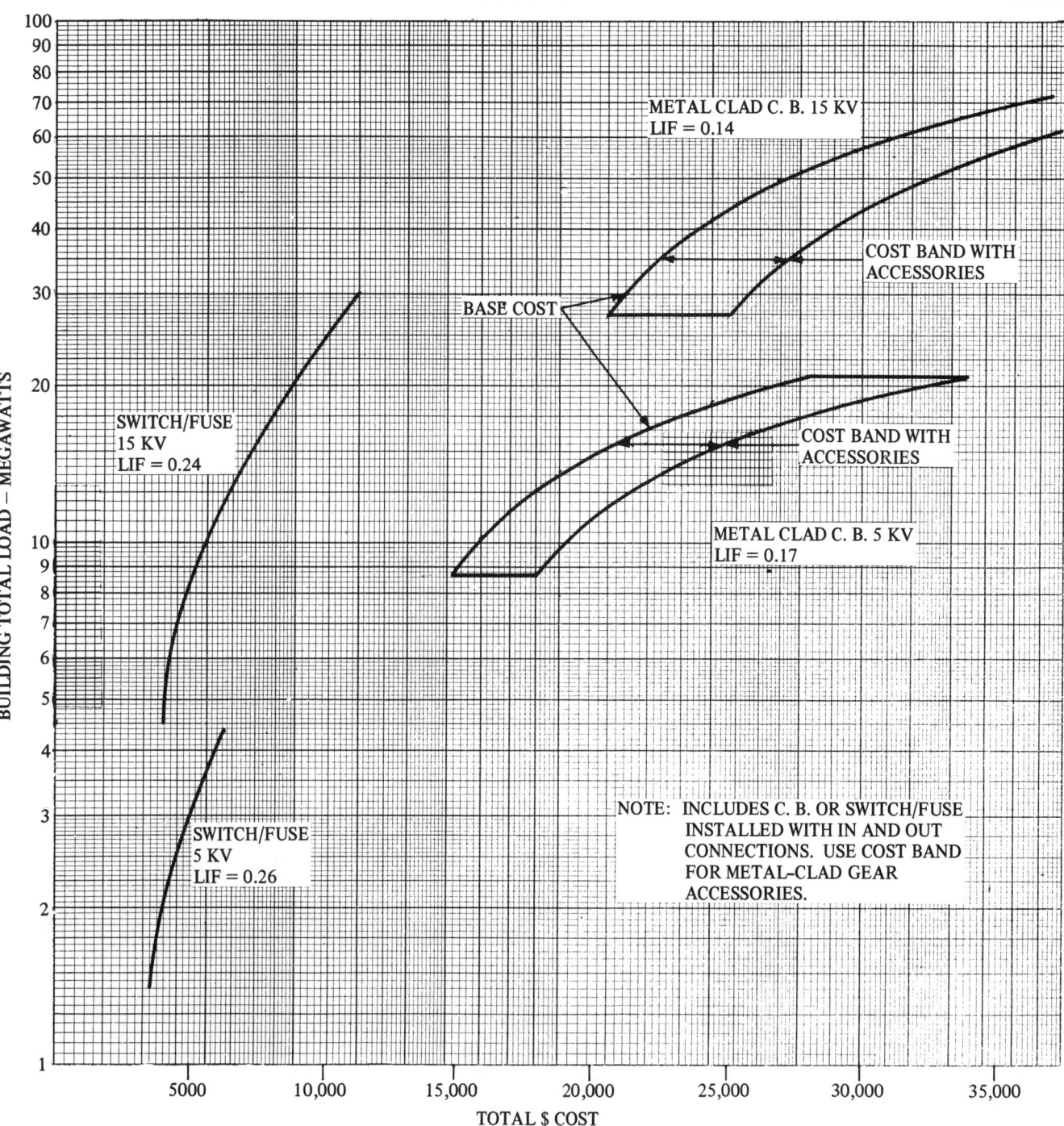

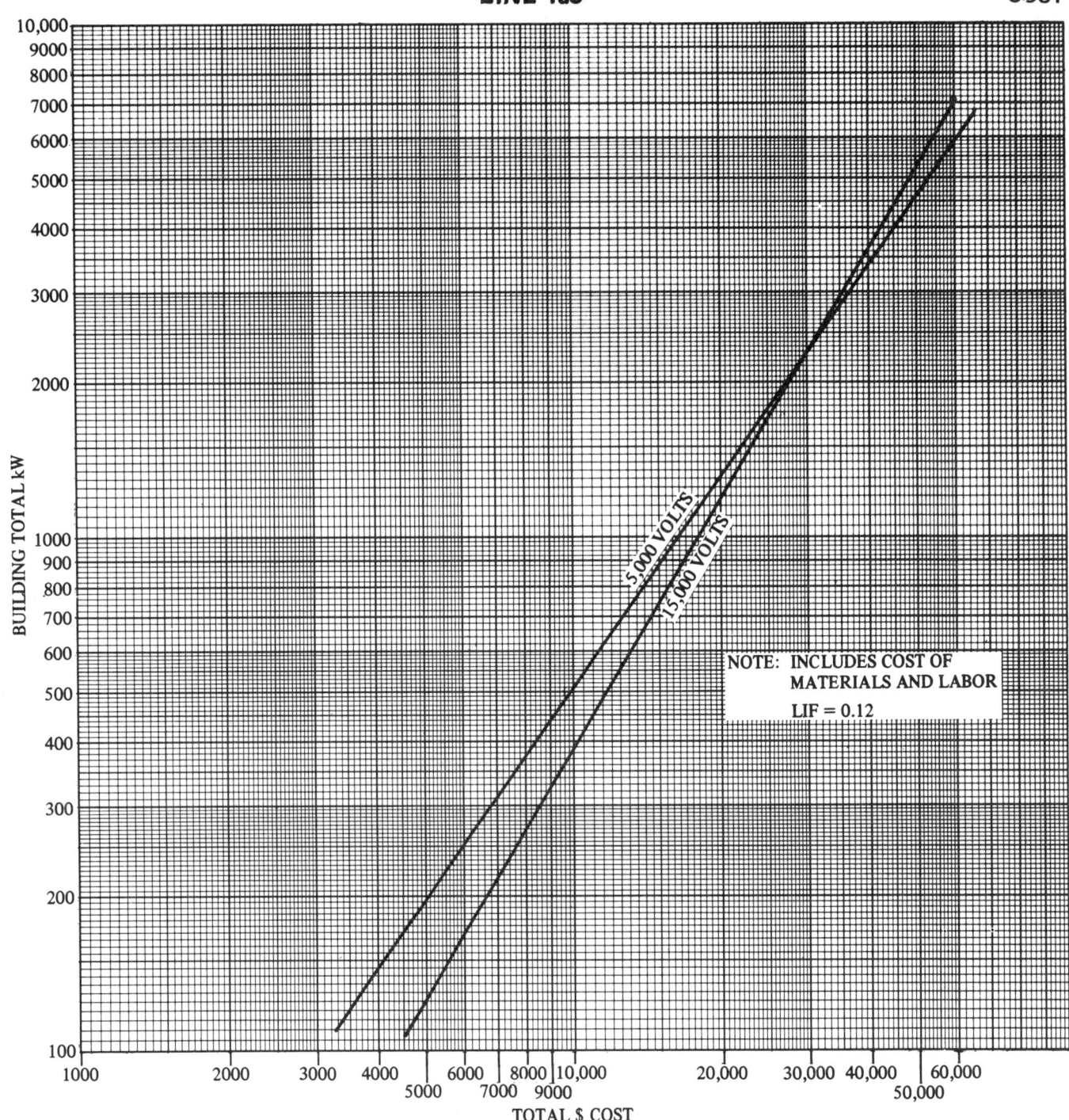

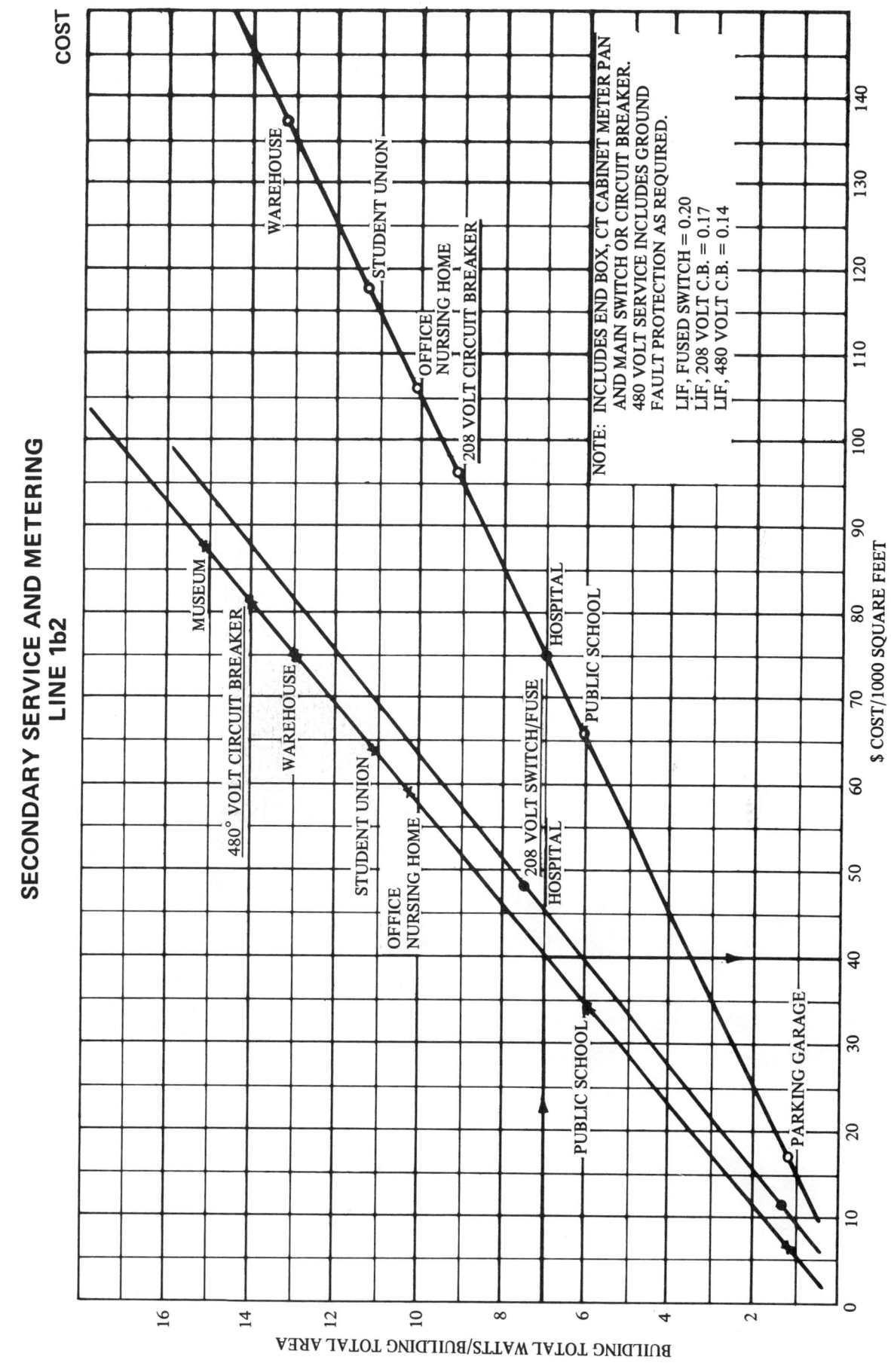

SECONDARY SERVICE AND METERING
LINE 1b2

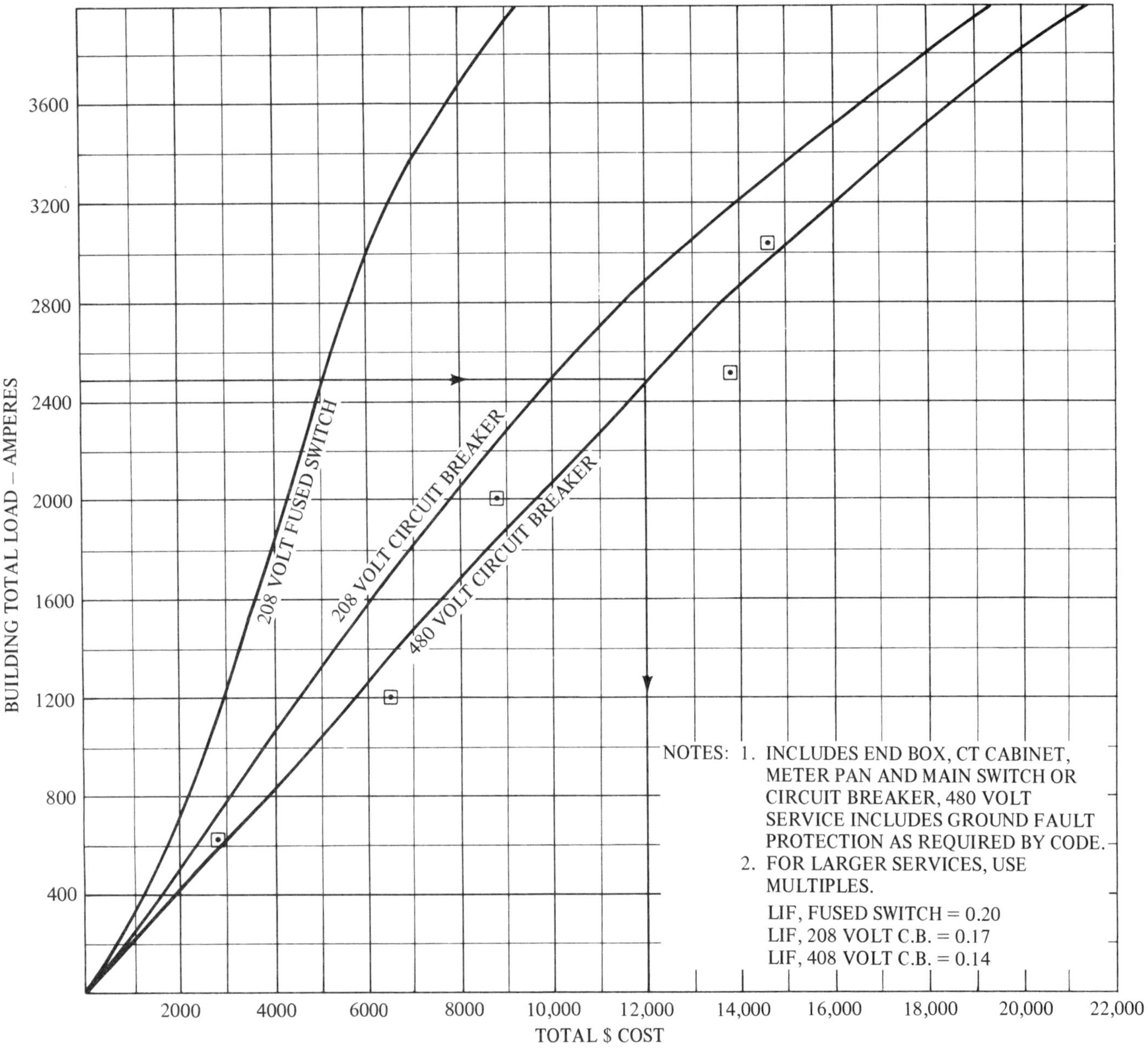

COST

NOTES:
1. INCLUDES END BOX, CT CABINET, METER PAN AND MAIN SWITCH OR CIRCUIT BREAKER, 480 VOLT SERVICE INCLUDES GROUND FAULT PROTECTION AS REQUIRED BY CODE.
2. FOR LARGER SERVICES, USE MULTIPLES.

LIF, FUSED SWITCH = 0.20
LIF, 208 VOLT C.B. = 0.17
LIF, 408 VOLT C.B. = 0.14

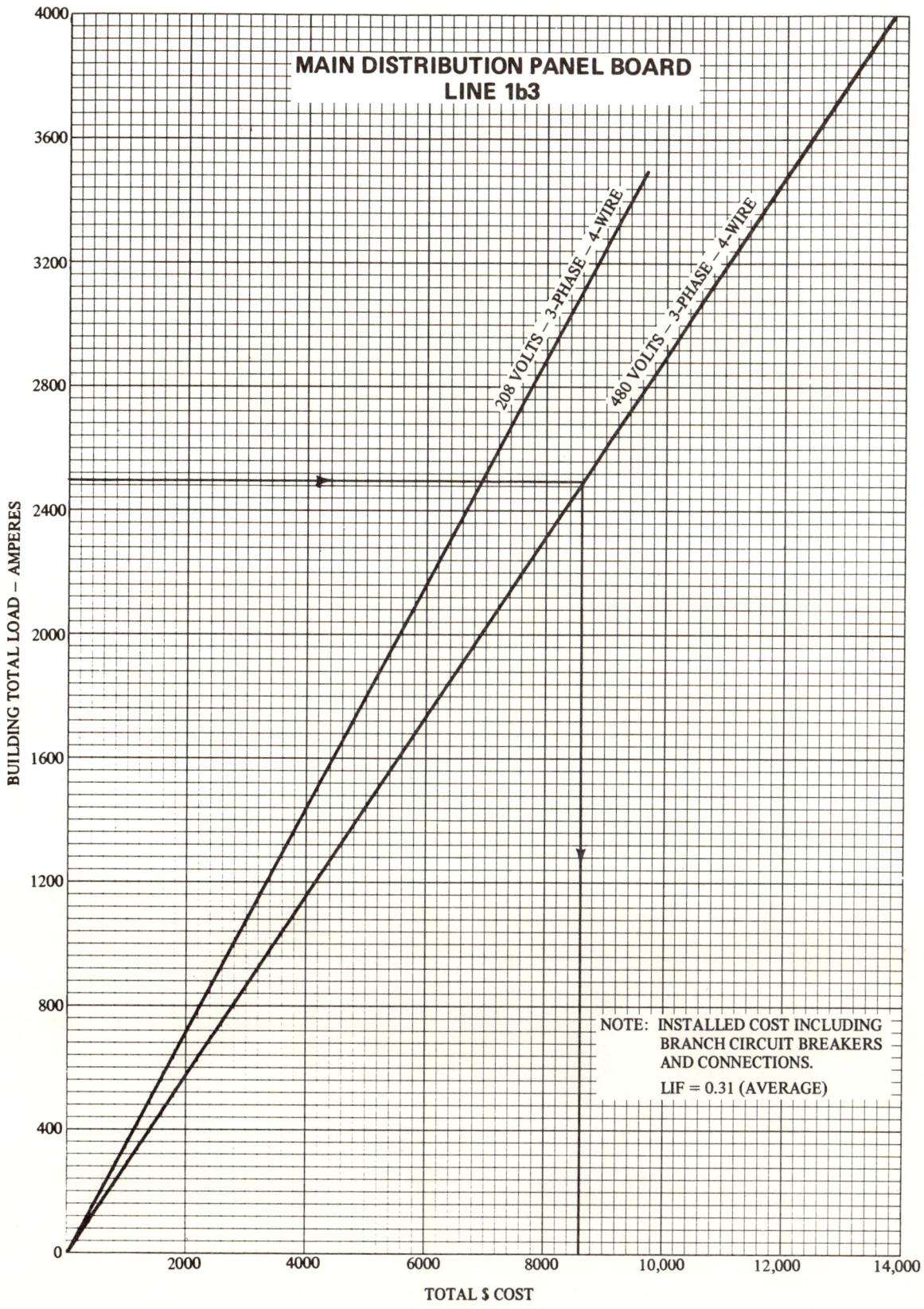

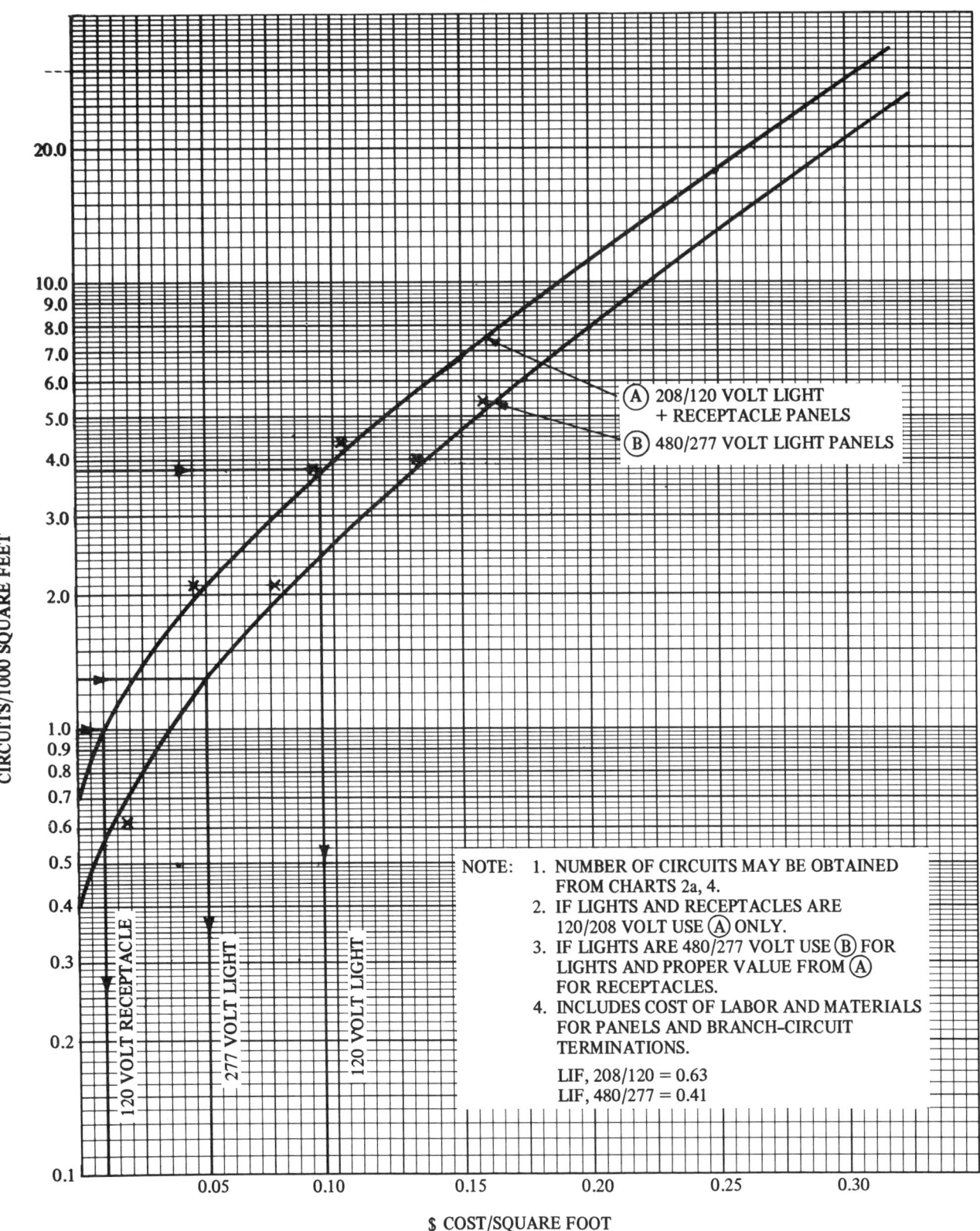

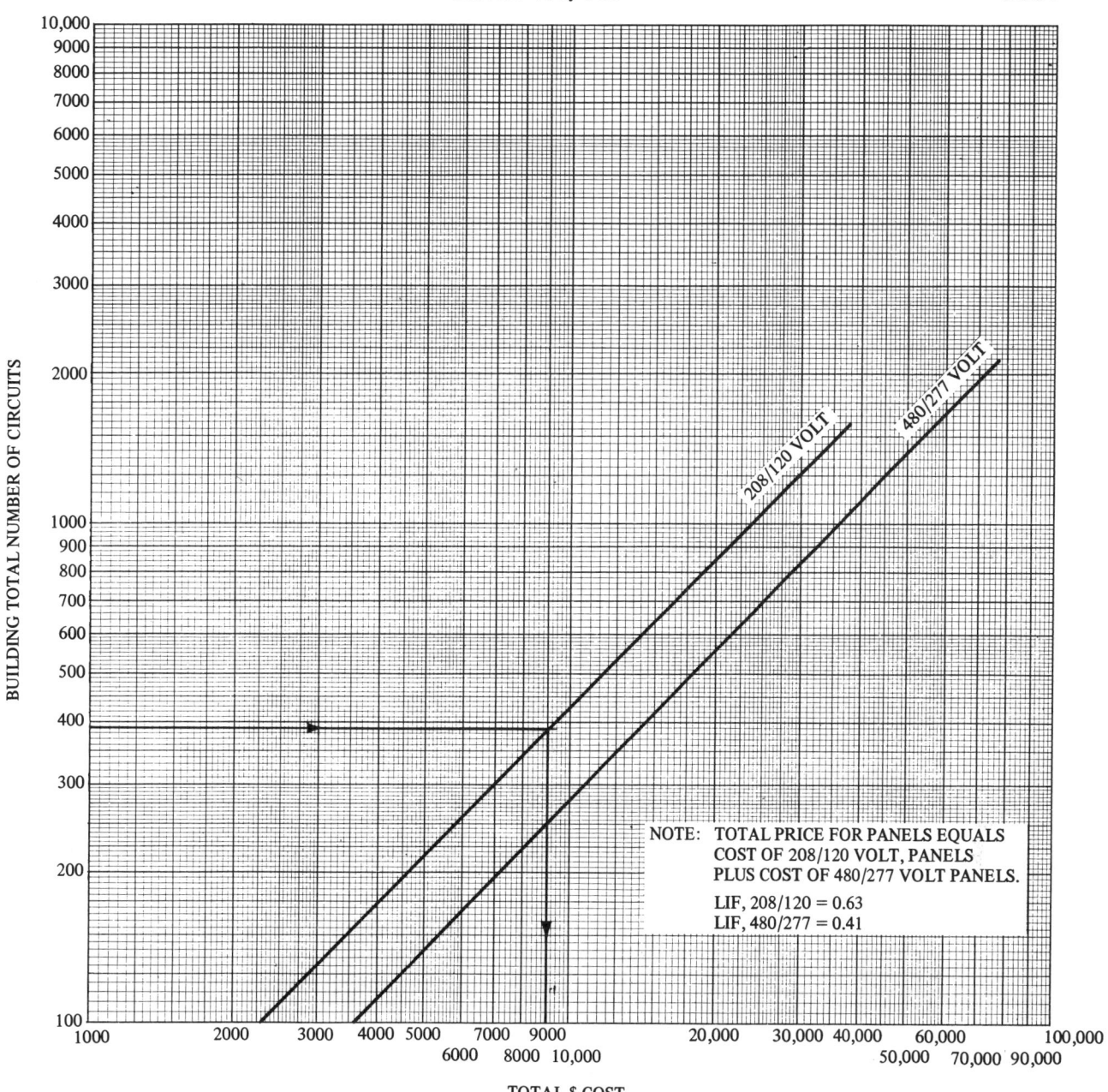

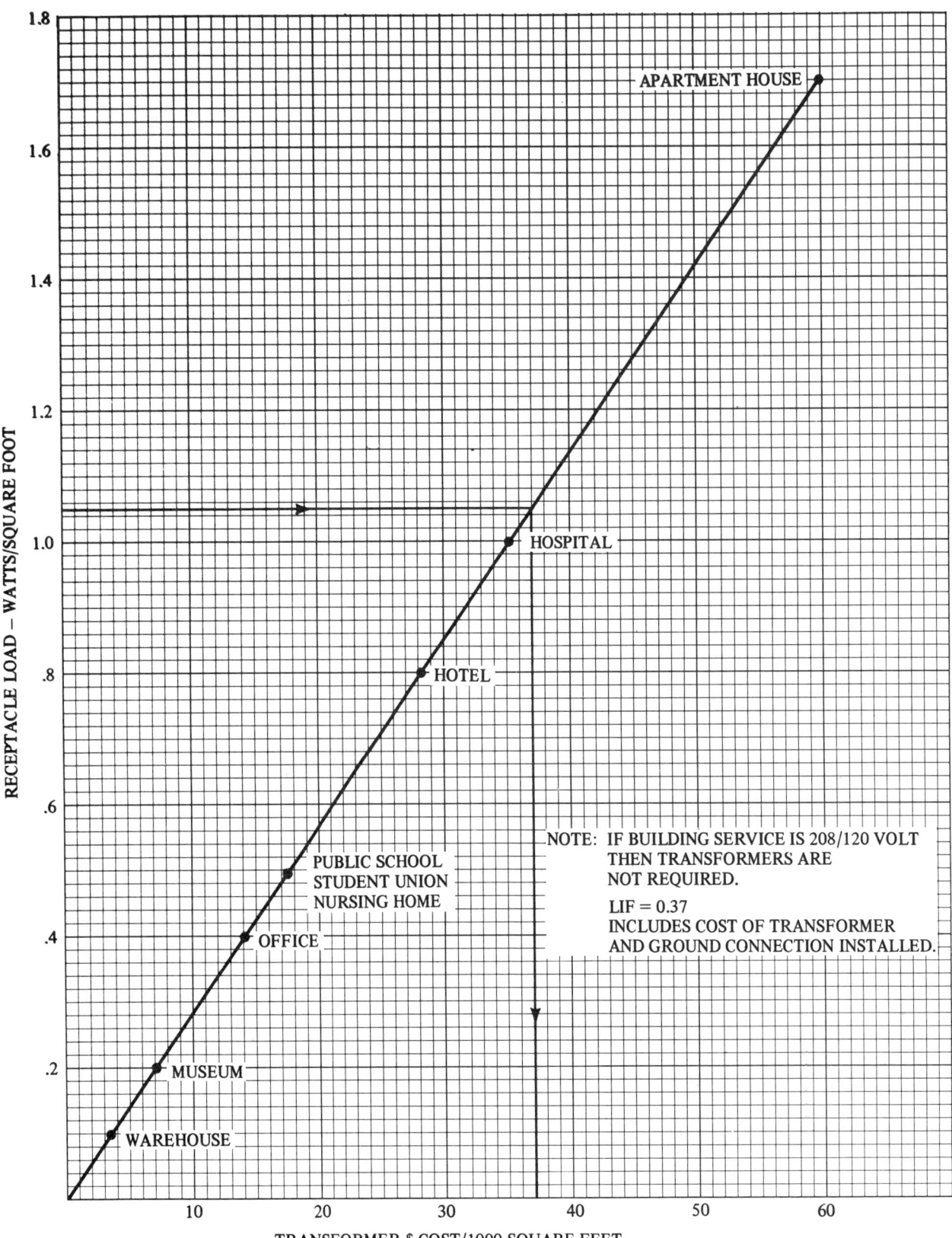

POWER PANELS
LINE 1c4

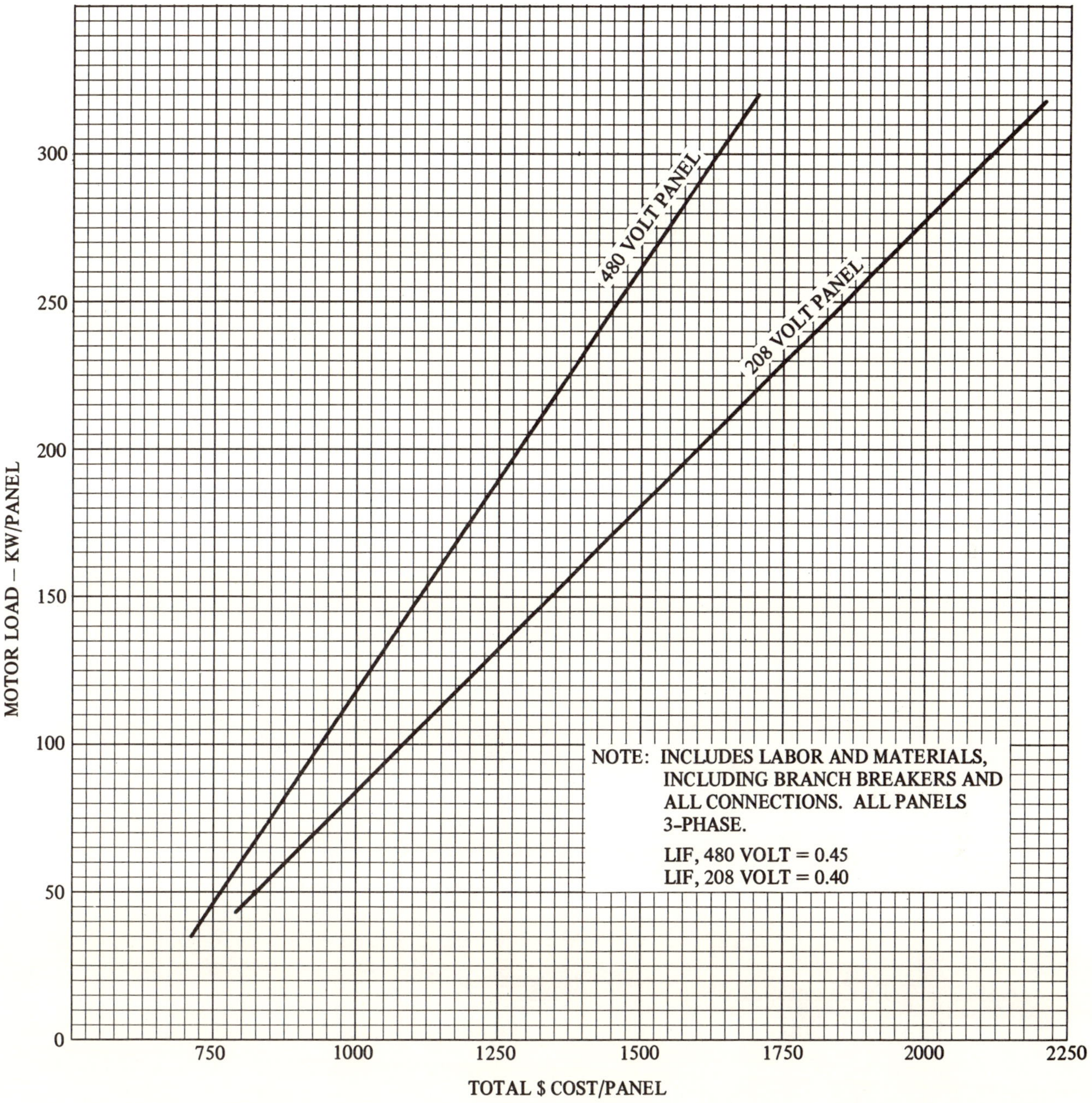
300
250
200
150
100
50
0
MOTOR LOAD – KW/PANEL
750
1000
1250
1500
1750
2000
2250
TOTAL $ COST/PANEL
480 VOLT PANEL
208 VOLT PANEL
NOTE: INCLUDES LABOR AND MATERIALS, INCLUDING BRANCH BREAKERS AND ALL CONNECTIONS. ALL PANELS 3-PHASE.
LIF, 480 VOLT = 0.45
LIF, 208 VOLT = 0.40

MOTOR FEEDERS
LINE 1d1

COST

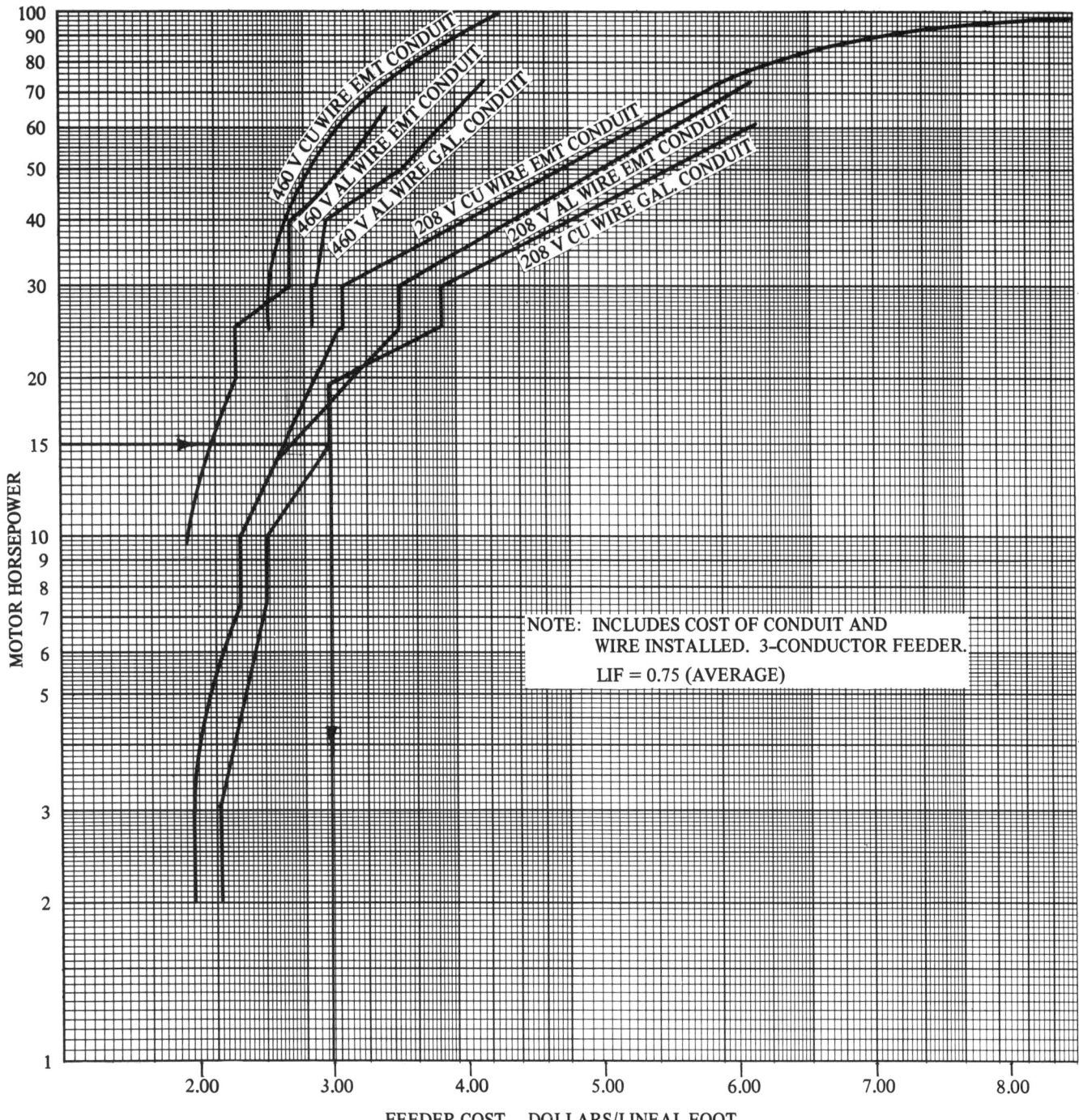

NOTE: INCLUDES COST OF CONDUIT AND WIRE INSTALLED. 3-CONDUCTOR FEEDER.
LIF = 0.75 (AVERAGE)

COST

3-PHASE, 3-WIRE – 600 VOLT THW INSULATION
LINES 1d1; 1d2

NOTES:
1. INCLUDES COST OF WIRE AND CONDUIT INSTALLED.
2. INTERMEDIATE METAL CONDUIT IS ABOUT THE SAME COST AS ALUMINUM CONDUIT.

LIF = 0.76 AVERAGE

RIGID GALVANIZED CONDUIT
ALUMINUM CONDUIT
EMT CONDUIT

$ COST/1000 AMPERE FEET

AMPACITY

WIRE SIZE: #500 MCM, #400 MCM, #250 MCM, #4/0, #2/0, #2, #4, #6, #8, #10, #12

CHARTS 129

COPPER FEEDERS
3-PHASE, 3-WIRE — 600 VOLT THW INSULATION
LINES 1d1; 1d2

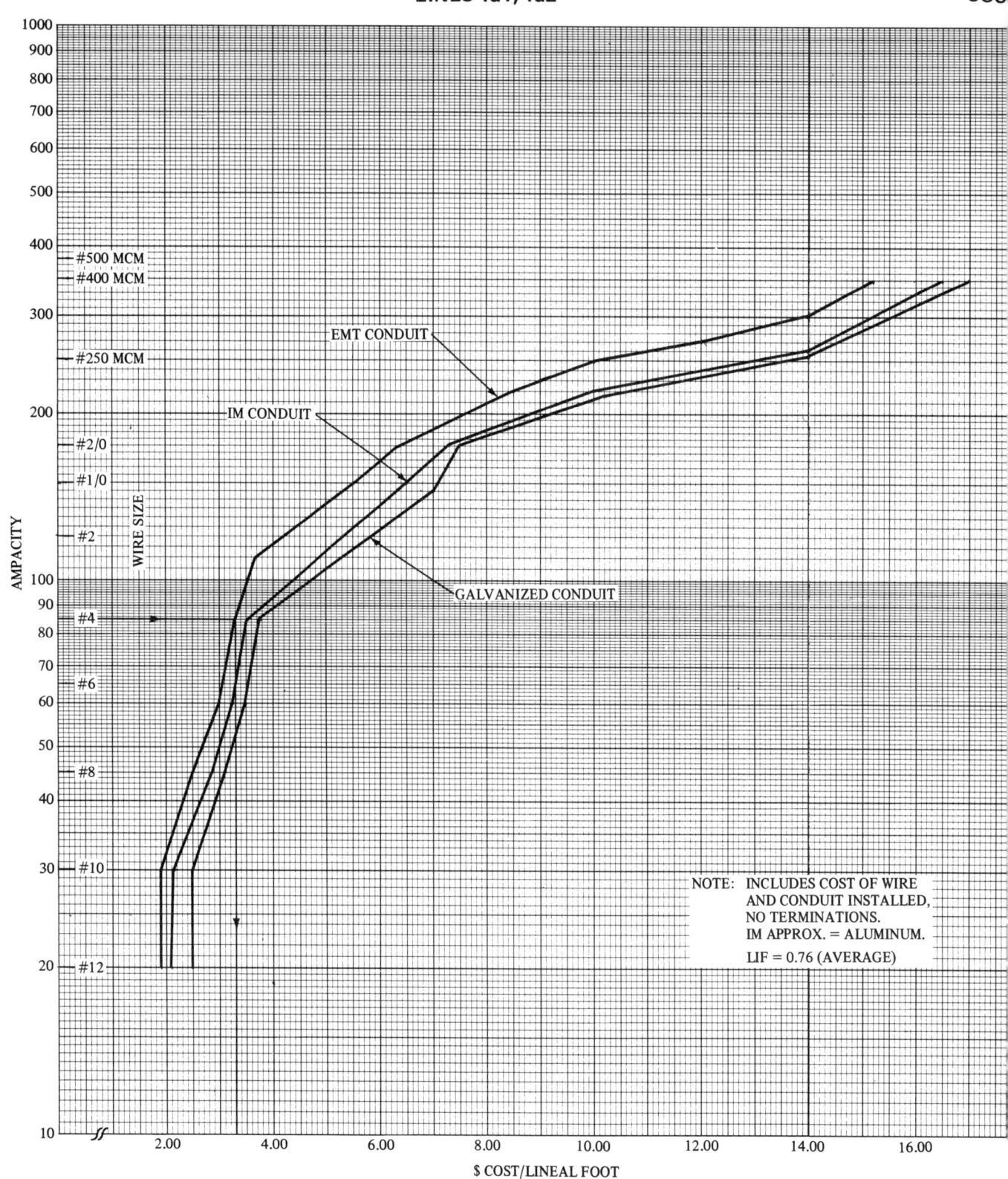

COPPER FEEDERS
3-PHASE, 4-WIRE — 600 VOLT THW INSULATION
LINES 1d1; 1d2

COST

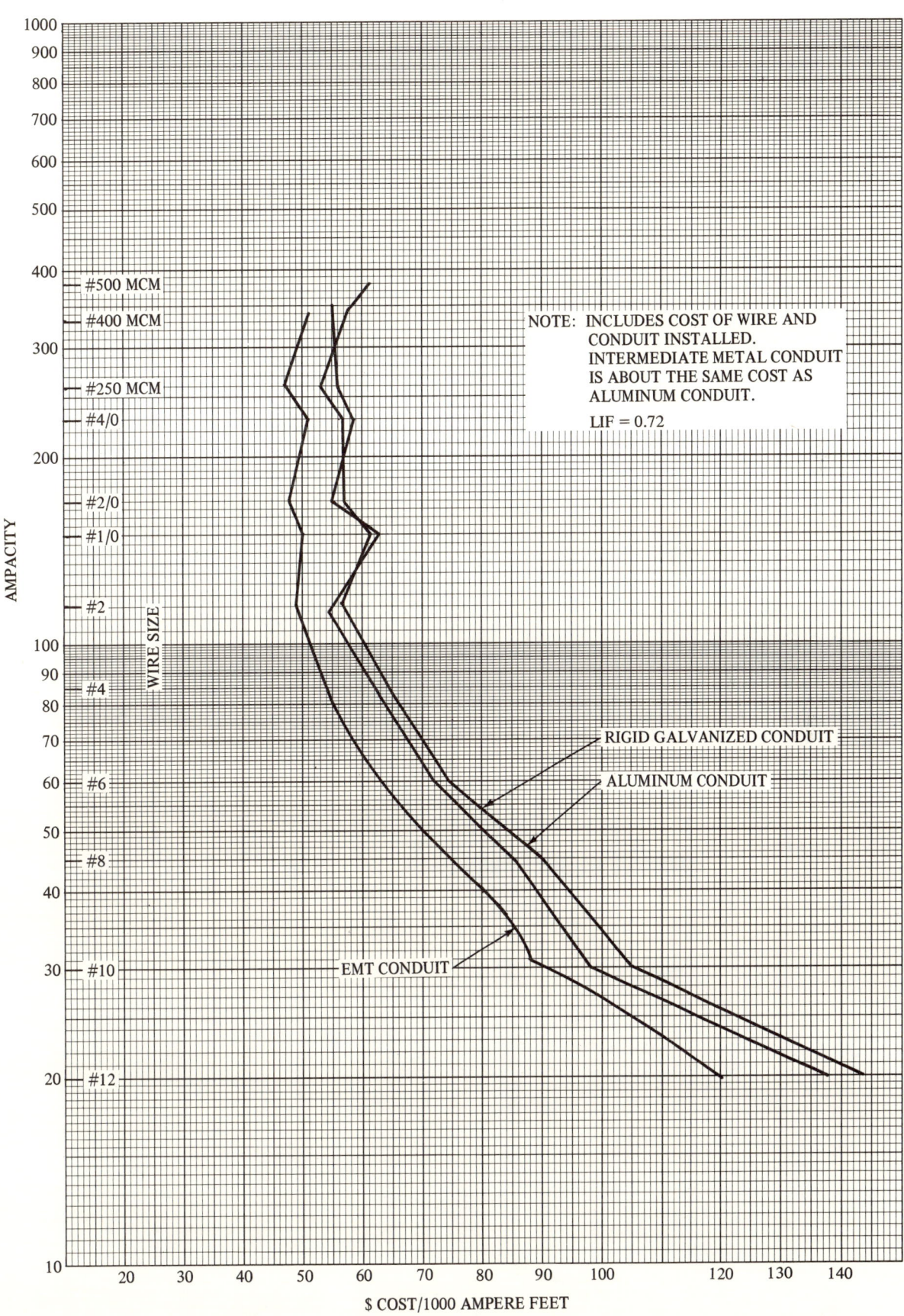

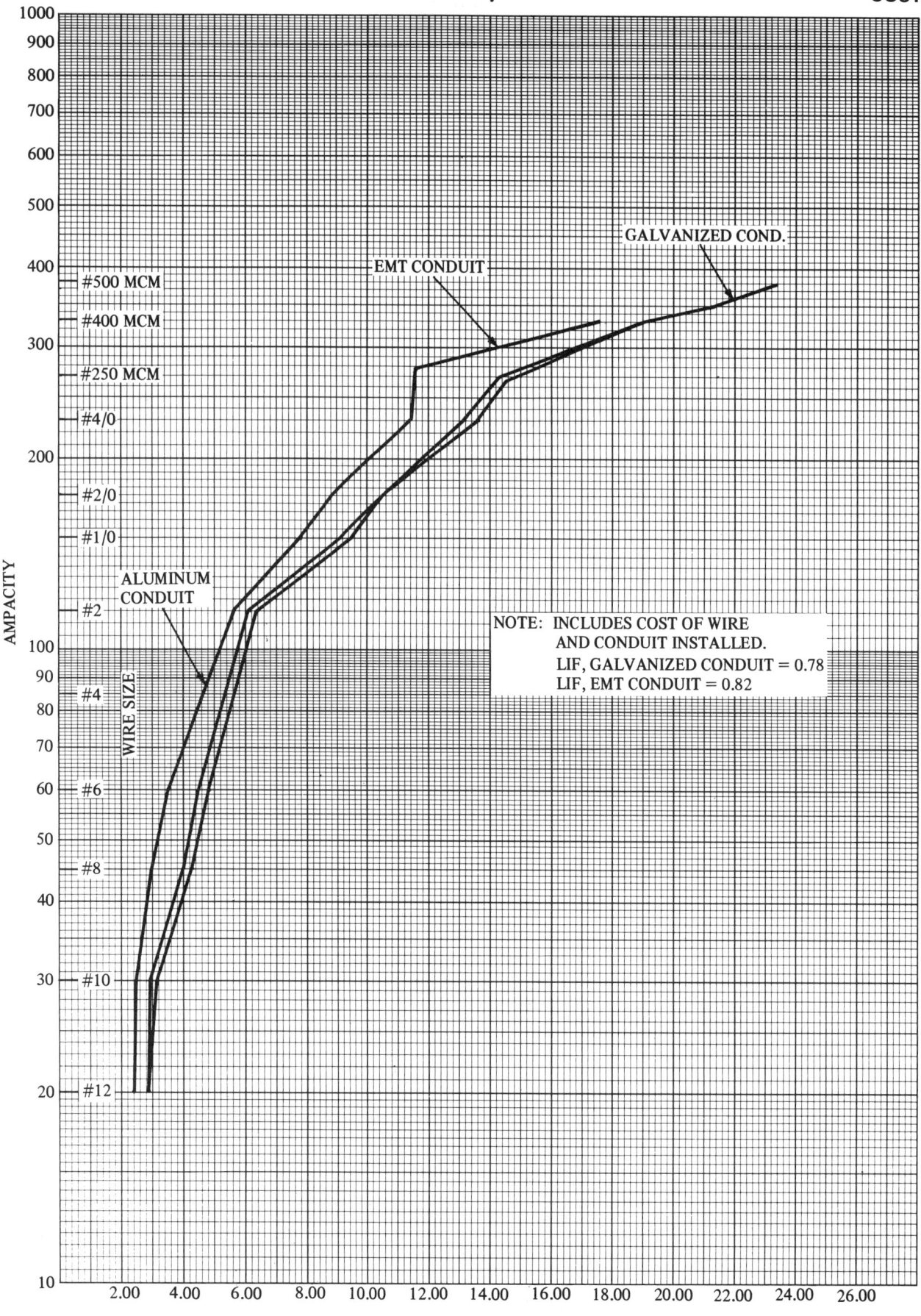

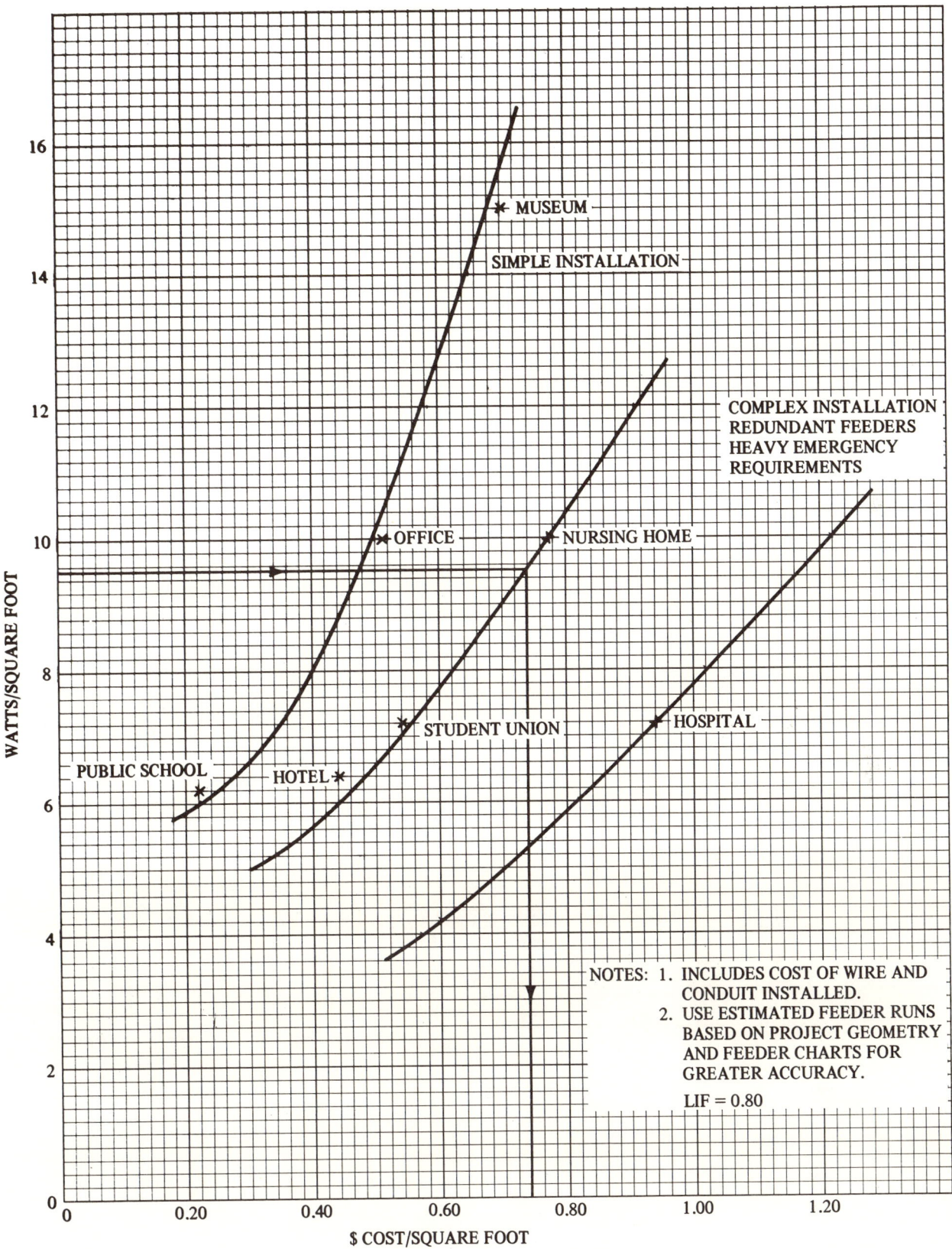

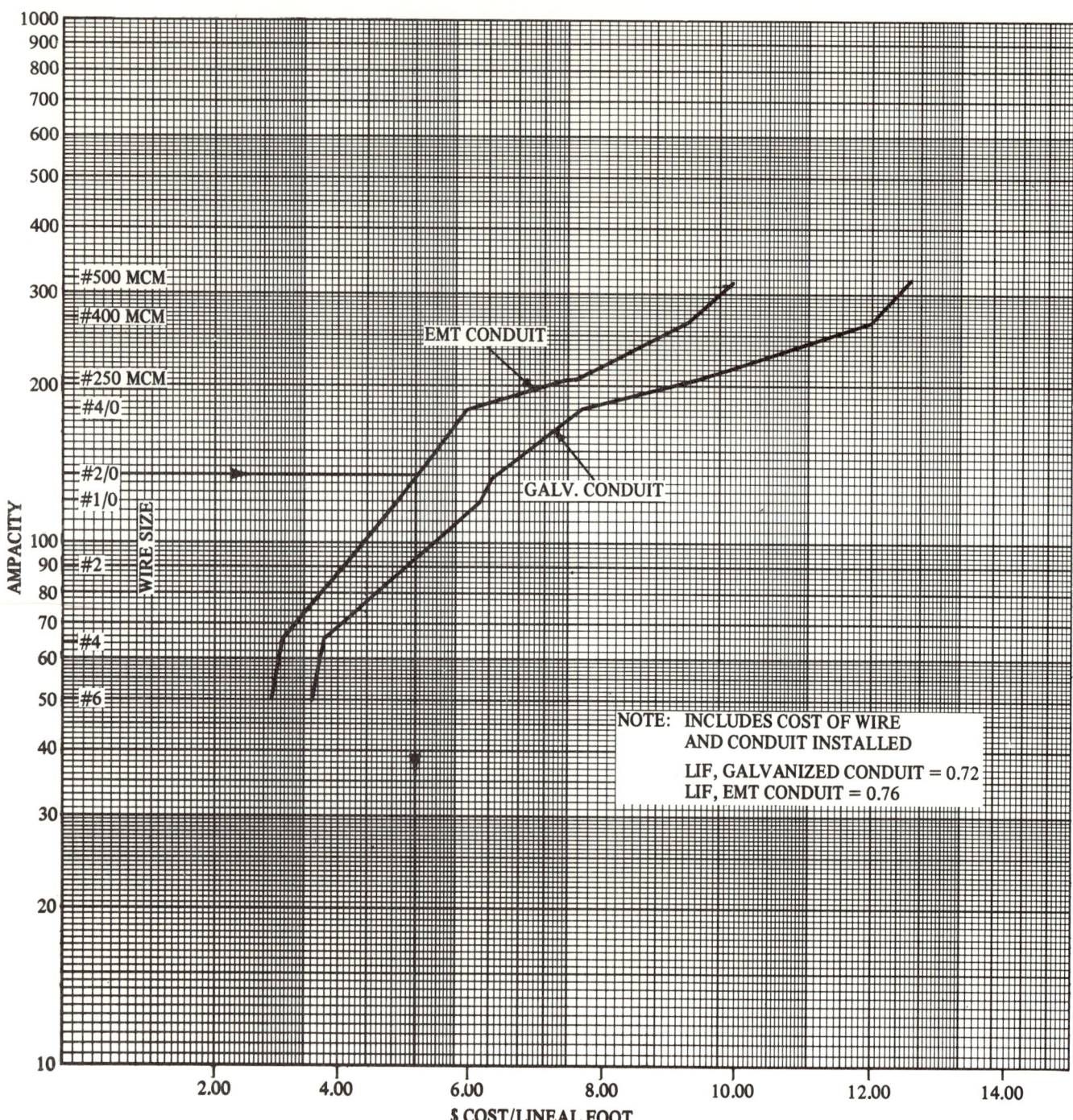

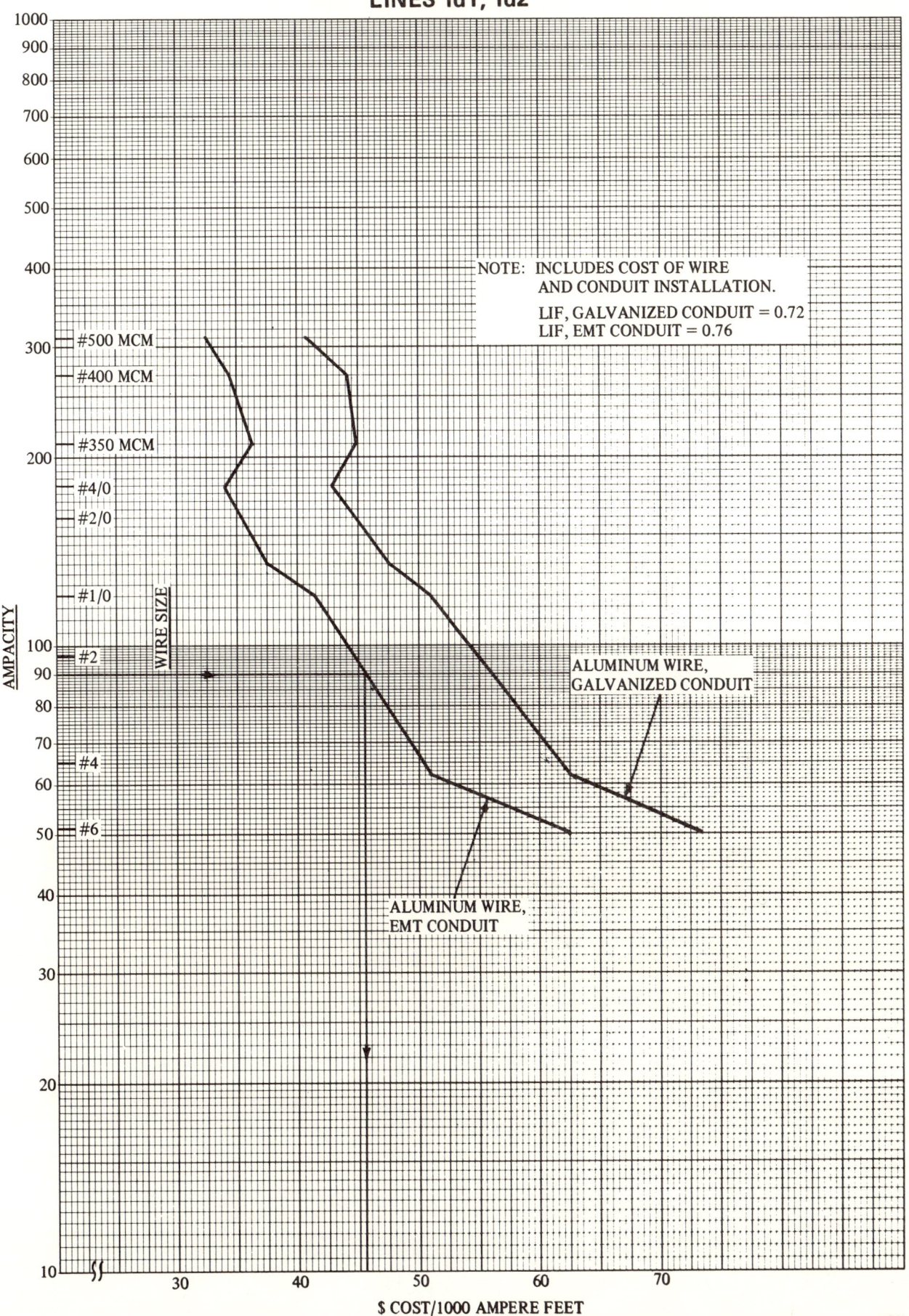

FEEDER BUS DUCT, 480 VOLT, 3-PHASE, 3-WIRE
LINES 1d1; 1d2

COST

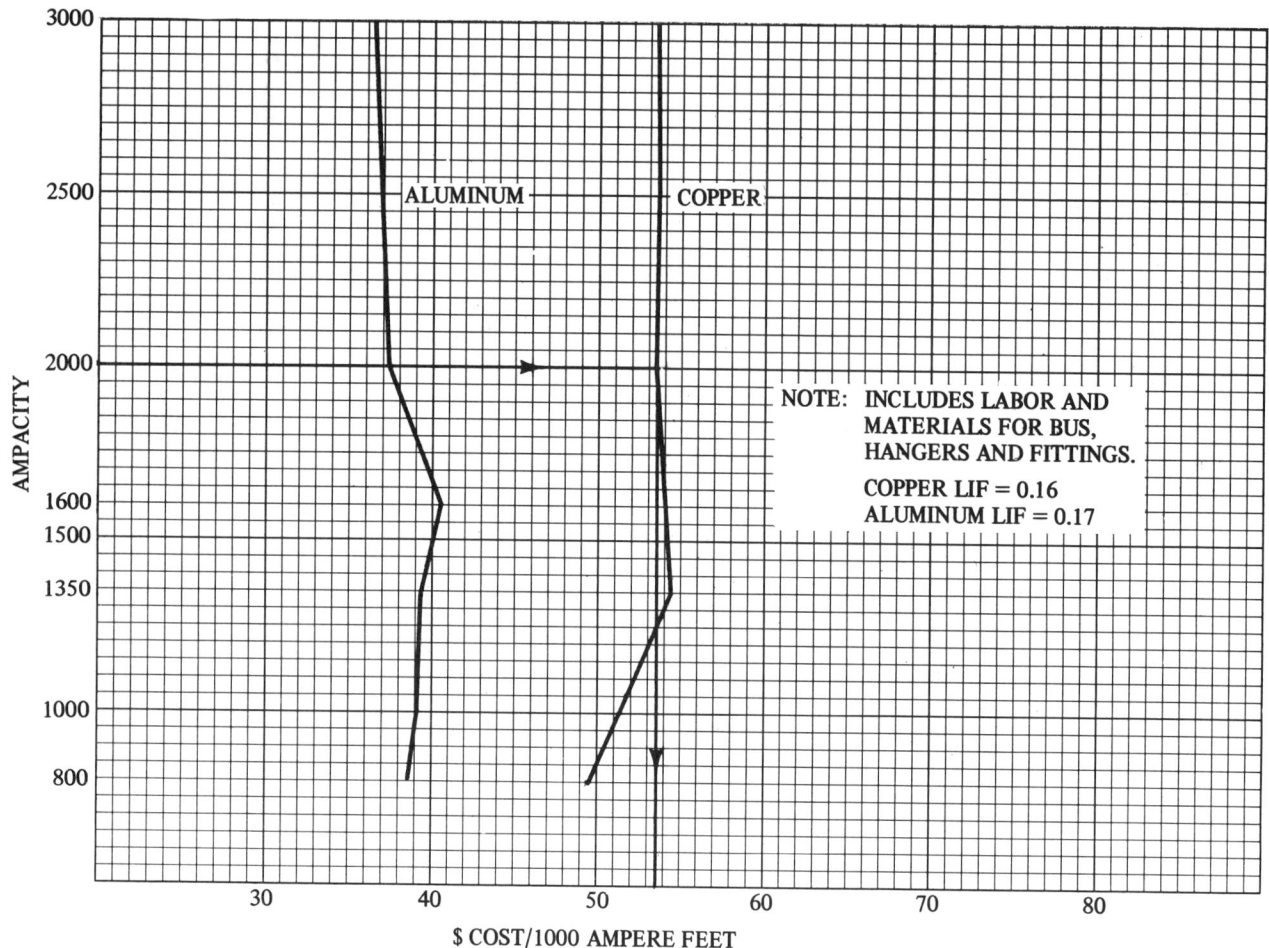

NOTE: INCLUDES LABOR AND MATERIALS FOR BUS, HANGERS AND FITTINGS.
COPPER LIF = 0.16
ALUMINUM LIF = 0.17

136 ESTIMATING AND COST CONTROL IN ELECTRICAL CONSTRUCTION DESIGN

EMERGENCY GENERATORS
LINE 1e

COST

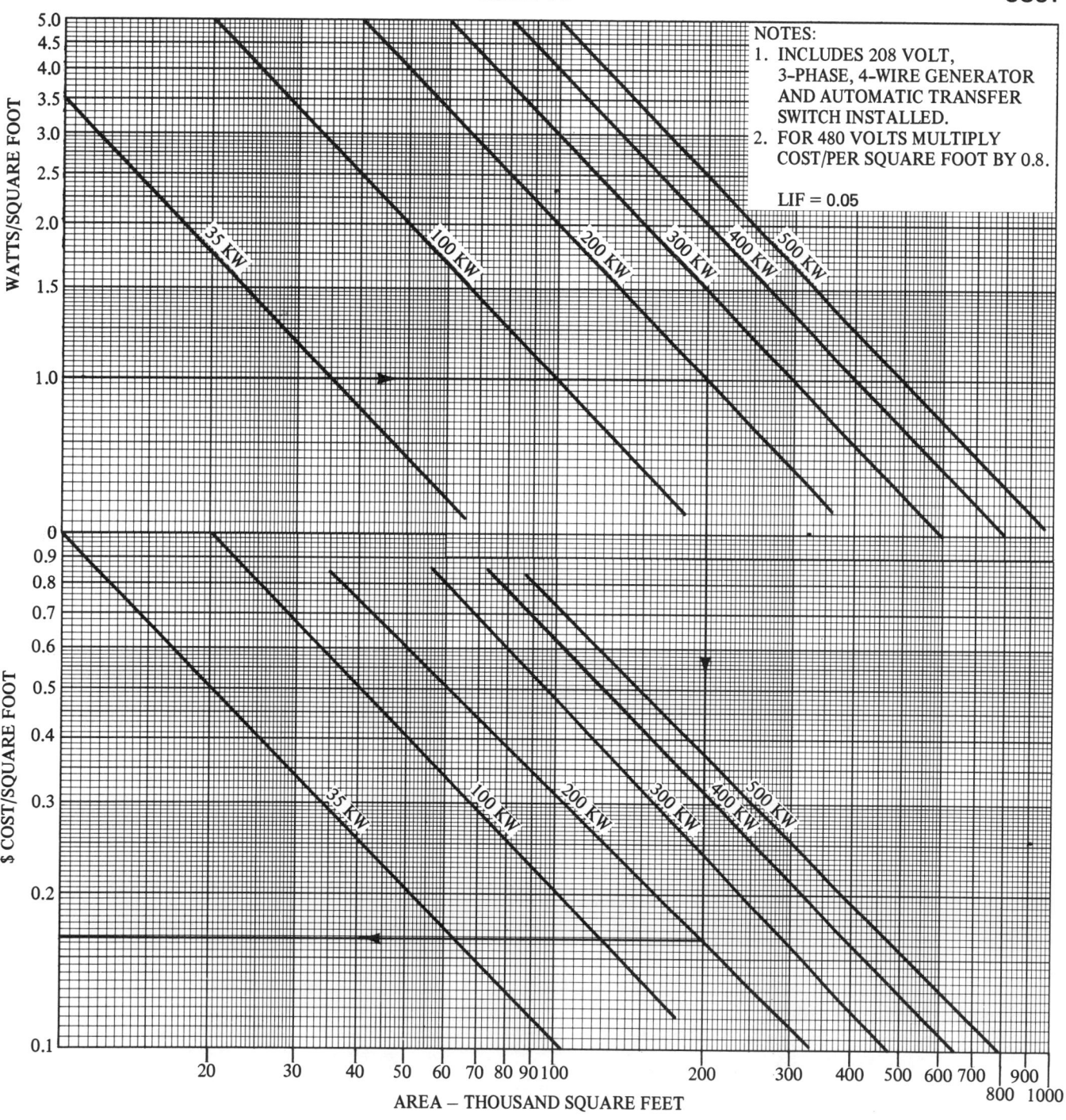

CHARTS 137

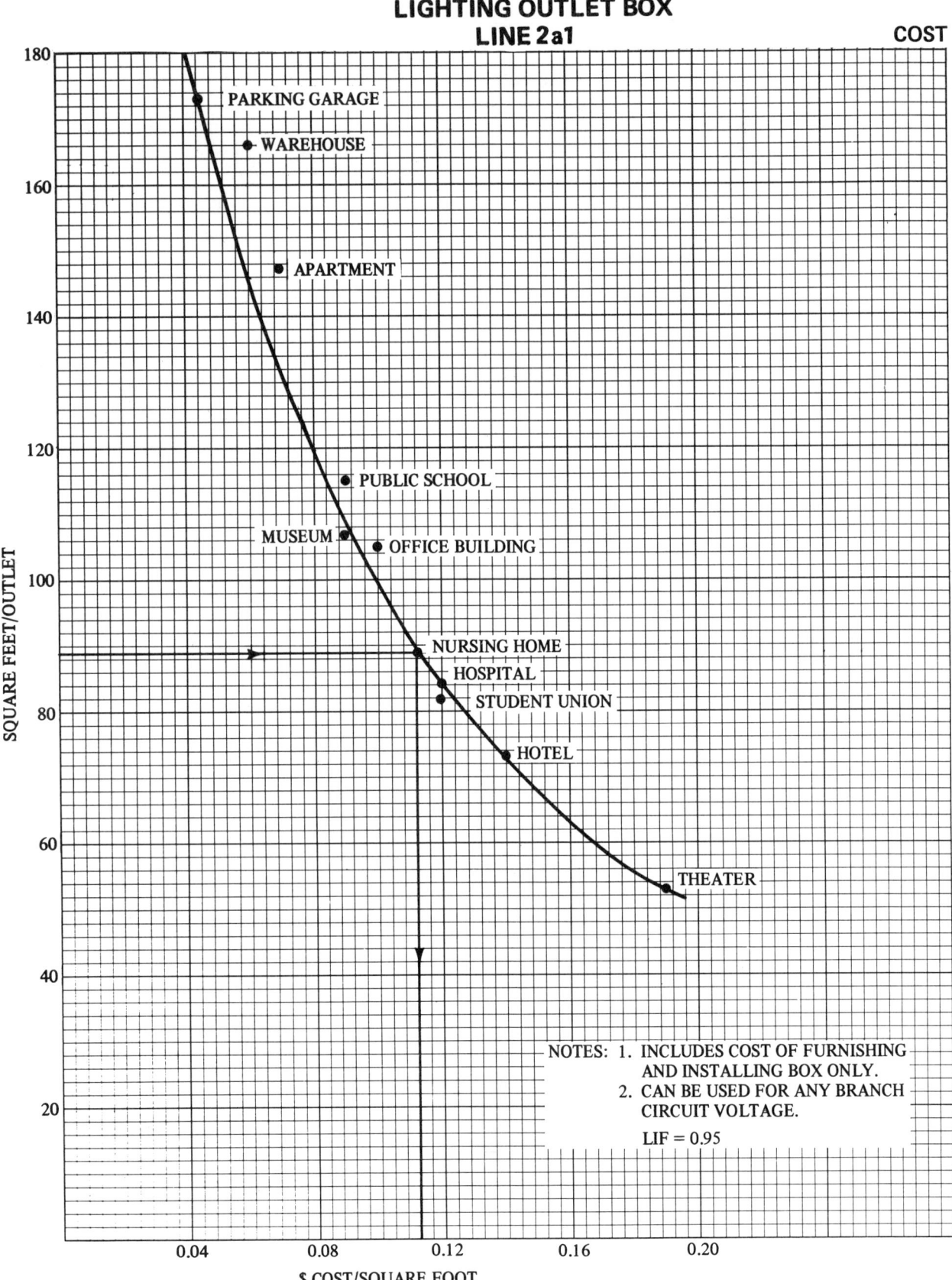

SWITCH AND RECEPTACLE
LINES 2a2; 2a3

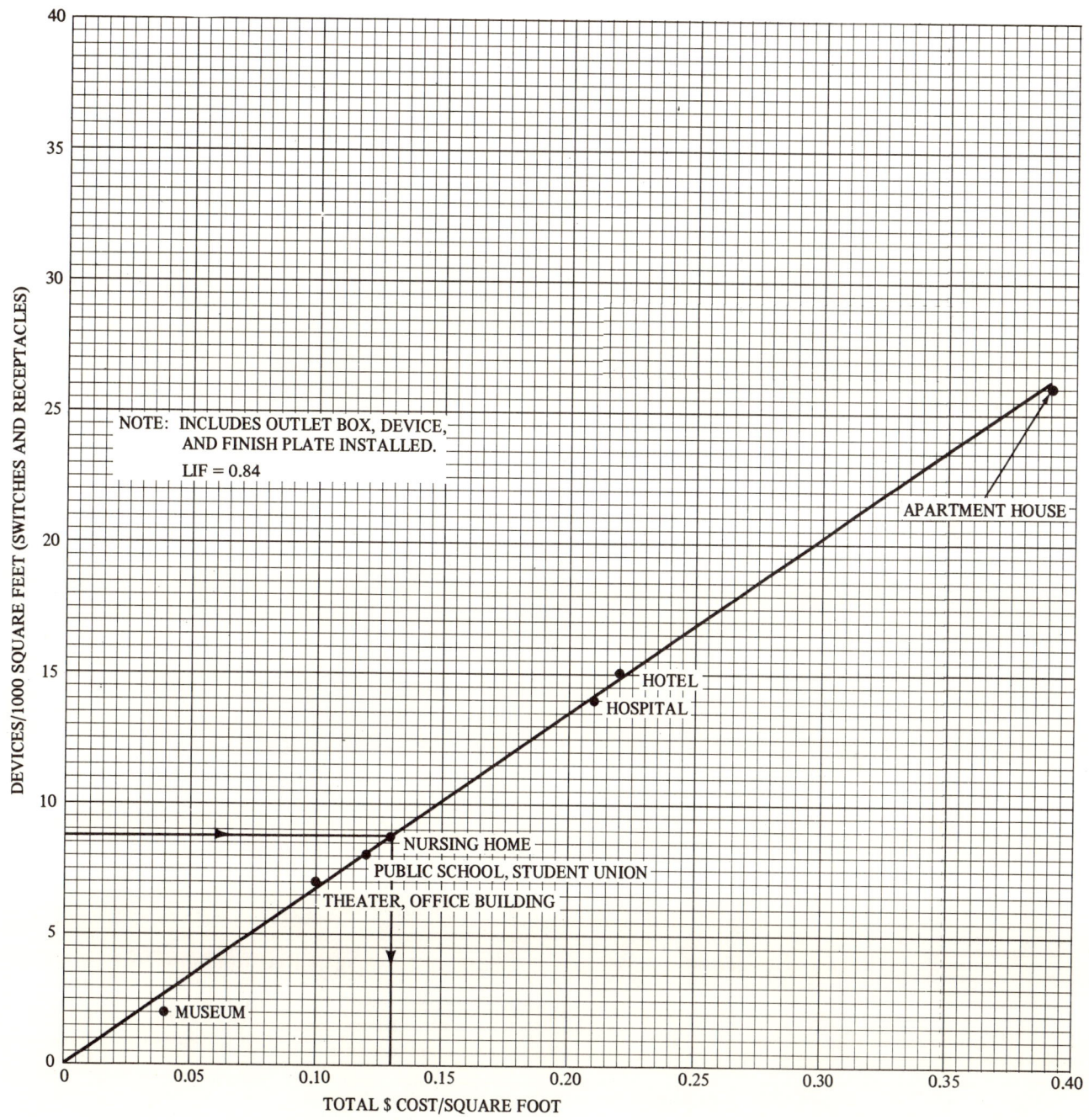

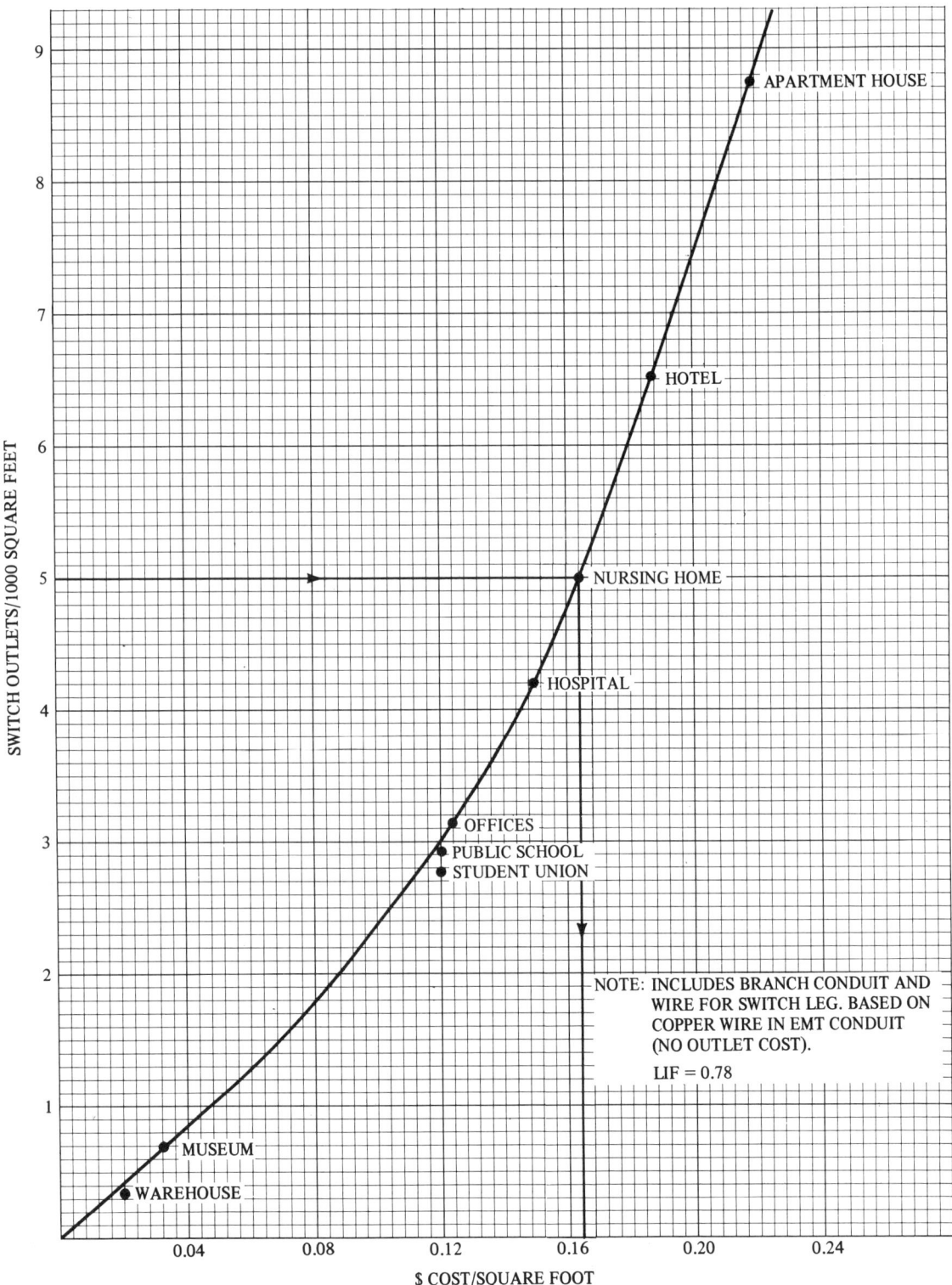

RECEPTACLE ELECTRICAL LOAD
LINE 2a4

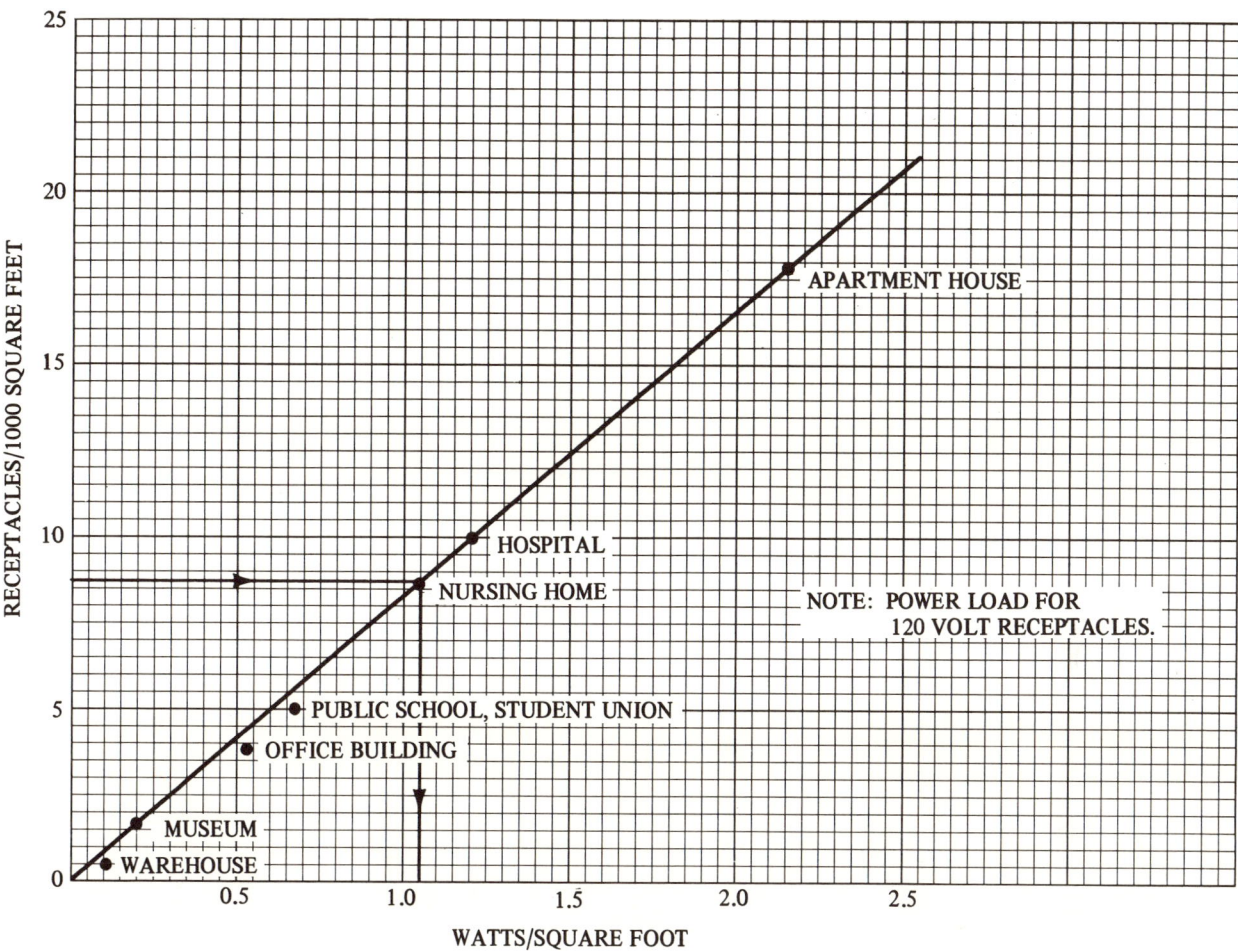

BRANCH-CIRCUIT WIRING, RECEPTACLES
LINE 2a4

COST

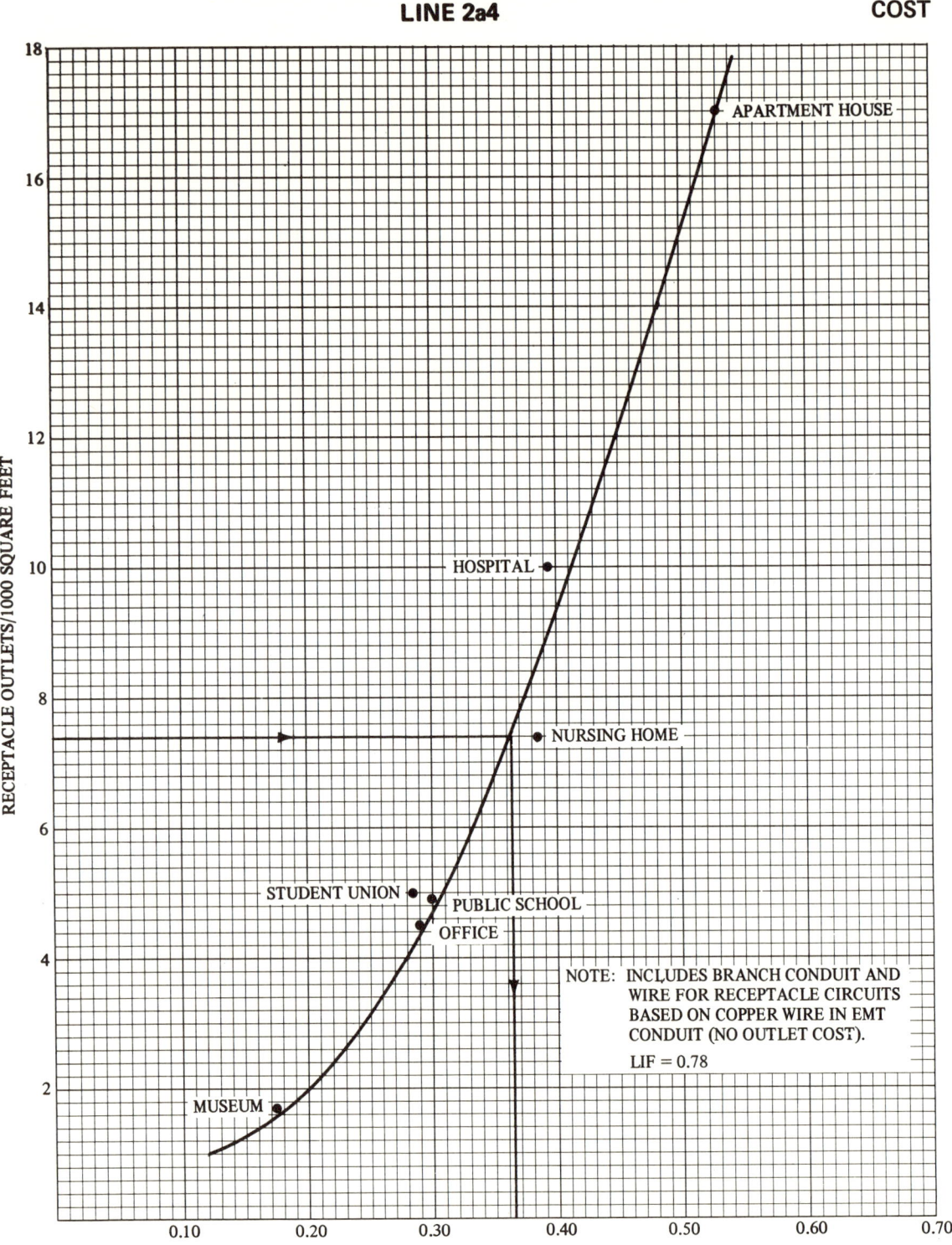

142 ESTIMATING AND COST CONTROL IN ELECTRICAL CONSTRUCTION DESIGN

BRANCH-CIRCUIT WIRING, LIGHTING
LINE 2a4

COST (vertical axis, top chart)

WATTS/SQUARE FOOT (vertical axis)

Curves labeled: 100, 200, 300, 400, 500, 600, 800, 1000 — **WATTS/OUTLET**

Building types indicated along curves: MUSEUM, OFFICE, PUBLIC SCHOOL, NURSING HOME, STUDENT UNION, HOSPITAL, HOTEL

Horizontal axis: **SQUARE FT/OUTLET** — 20, 40, 60, 80, 100, 120, 140, 160

OUTLET SPACING RATIO (lower chart)

Curves labeled: 1:1, 1:2, 1:3, 1:4, 1:5

Left-side scales:
- C/SF AT 6 WATTS/SF: .30, .40, .50, .60
- C/SF AT 4 WATTS/SF: .20, .30, .40, .50
- C/SF AT 2 WATTS/SF: .10, .20, .30, .40

Right-side scales:
- C/SF AT 10 WATTS/SF: 0.40, 0.50, 0.60, 0.70
- C/SF AT 8 WATTS/SF: 0.50, 0.60, 0.70, 0.80
- C/SF AT 12 WATTS/SF: 0.60, 0.70, 0.80, 0.90

NOTES:
1. INCLUDES BRANCH CONDUITS AND WIRE FOR LIGHTING OUTLETS BASED ON COPPER WIRE AND EMT CONDUIT. (NO OUTLET COST OR CONNECTIONS)
2. CHART IS DRAWN FOR 208/120 VOLTS, 3-PHASE. FOR 480/277 VOLTS DEDUCT $0.05 FOR EACH 2 W/SF. LIF = 0.78

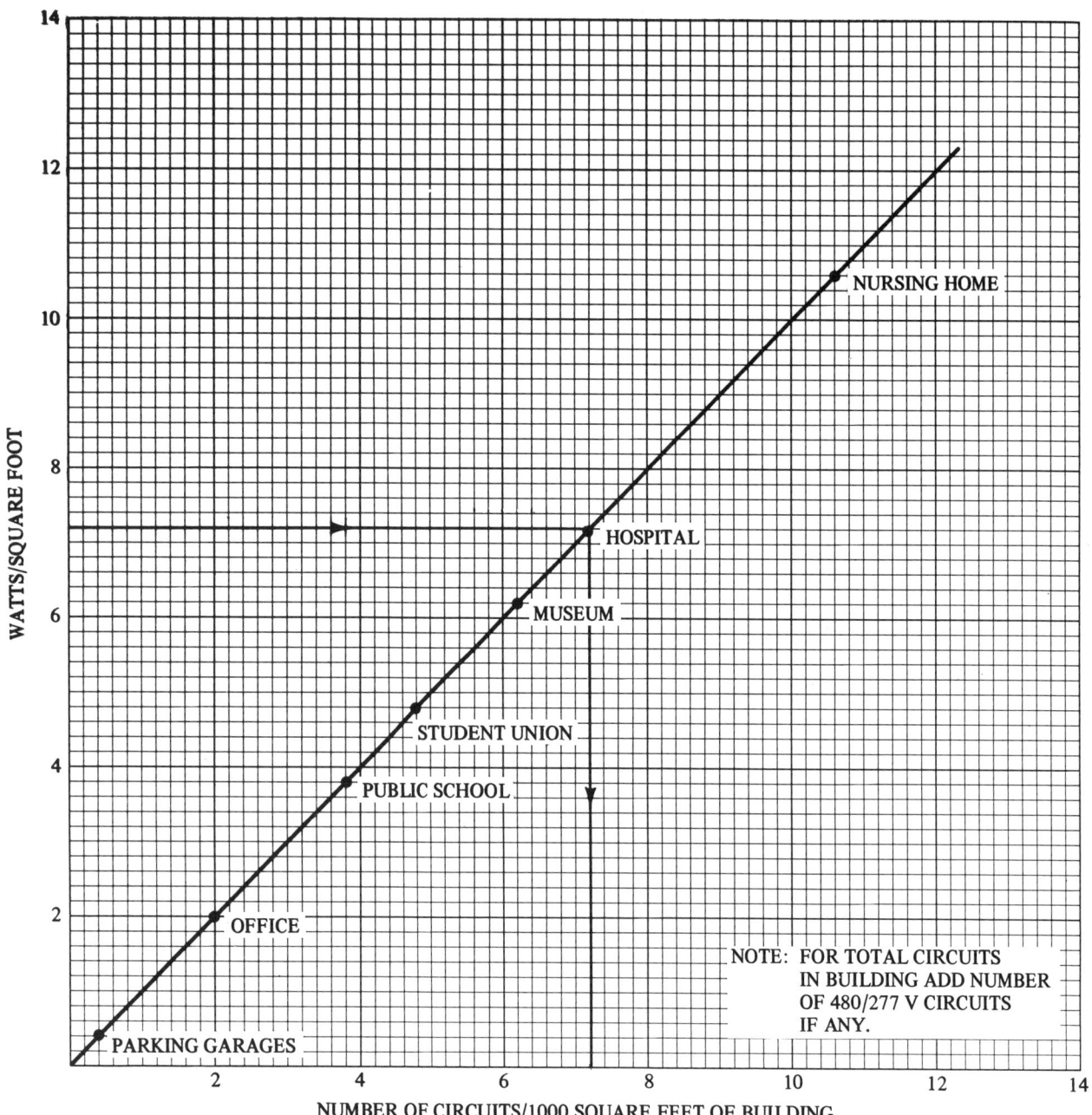

NUMBER OF BRANCH CIRCUITS
480/277 VOLT CIRCUITS, 277 V LIGHTING ONLY
LINE 2a4

LOAD

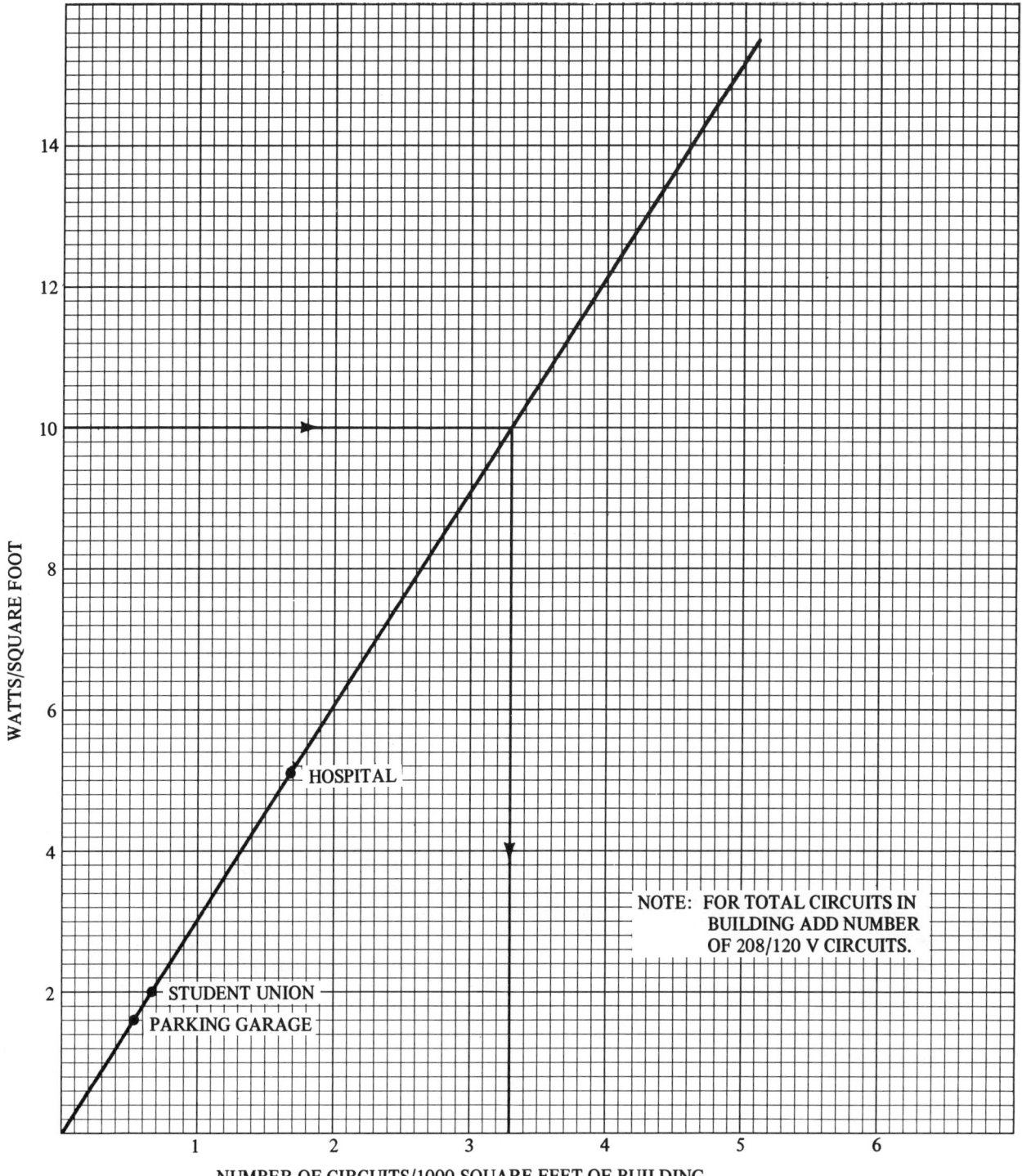

CHARTS 145

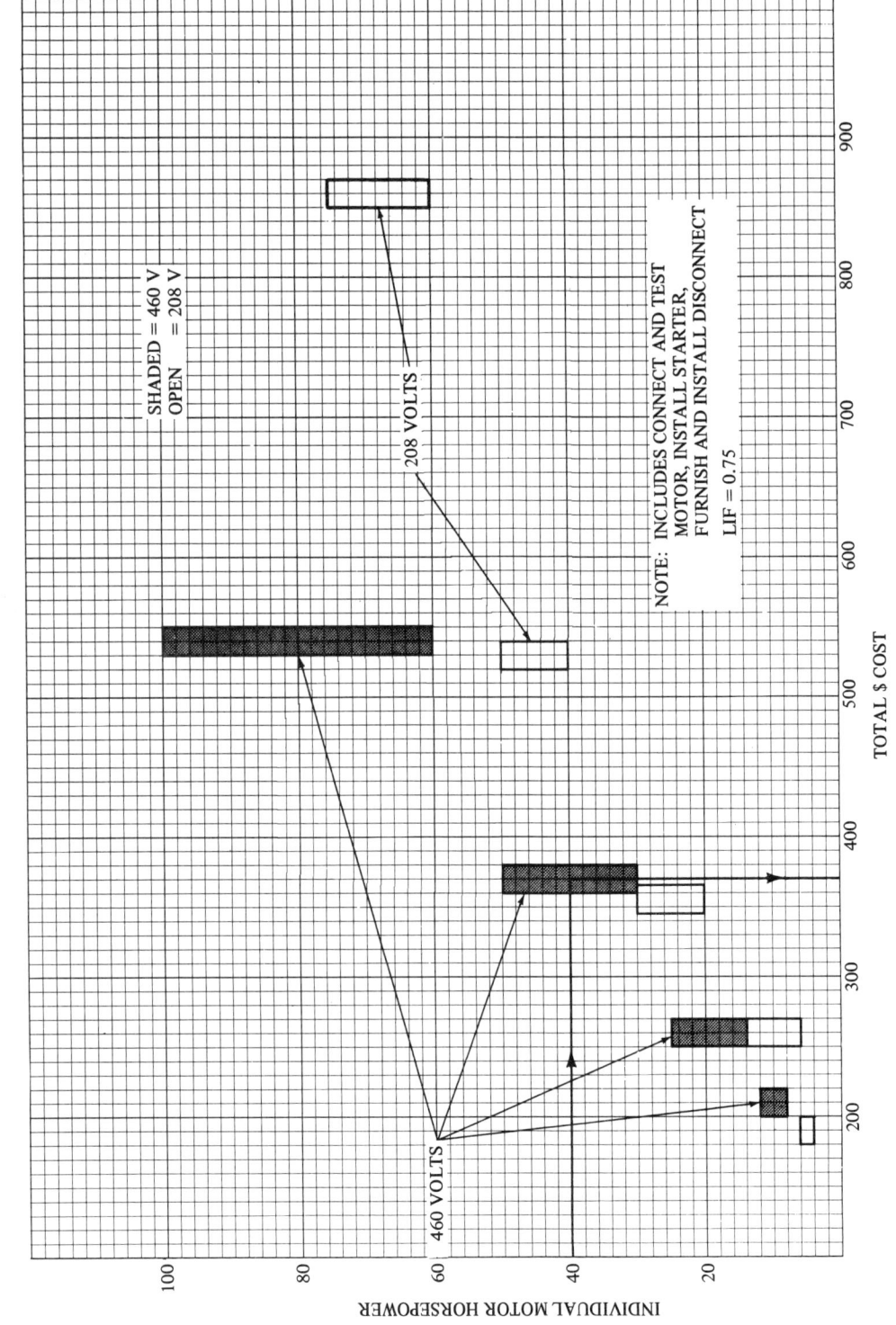

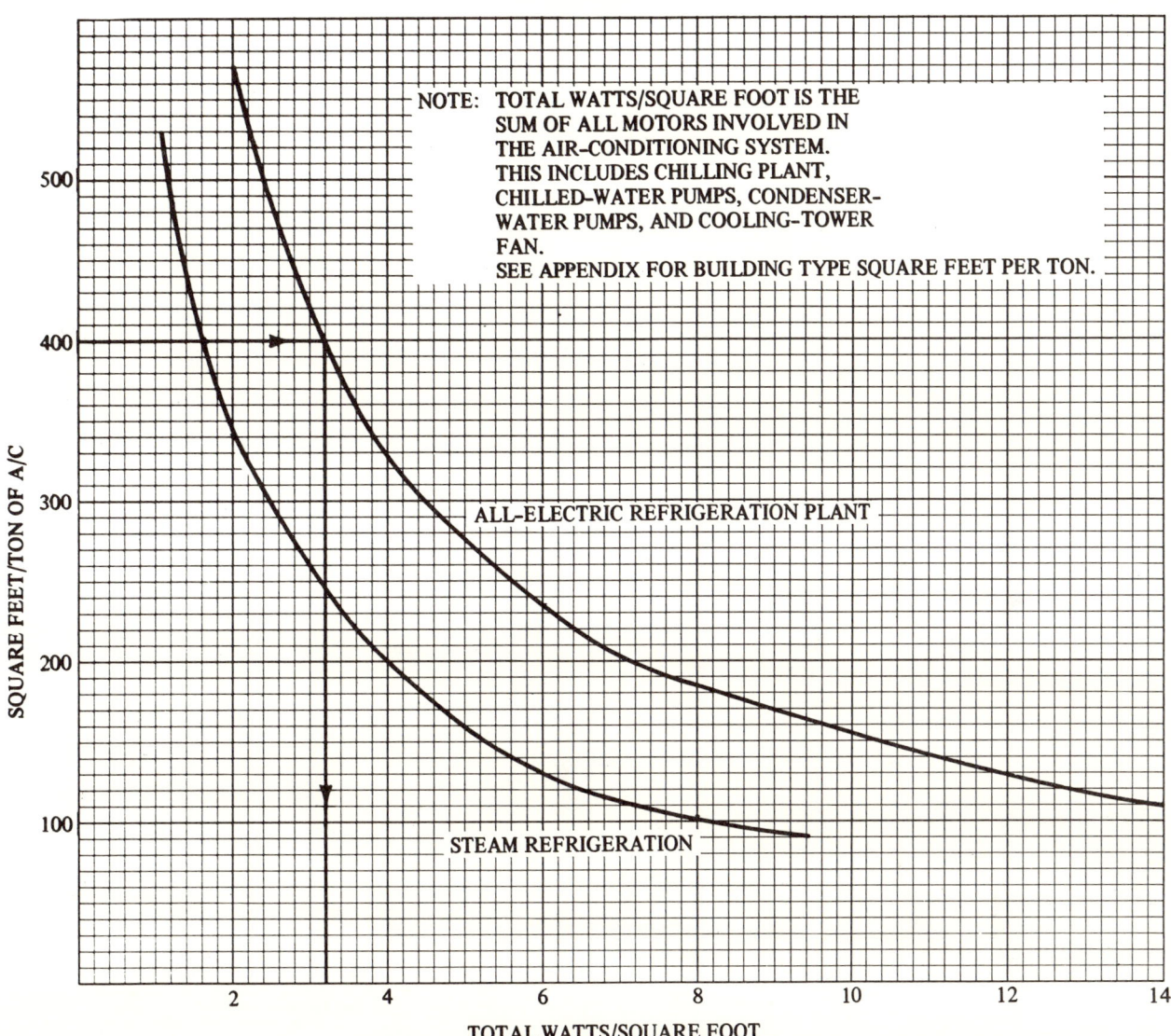

AIR CONDITIONING MOTORS AND CONTROL
(STEAM OR UTILITY CENTRAL COOLING PLANT)
LINE 2b2

COST

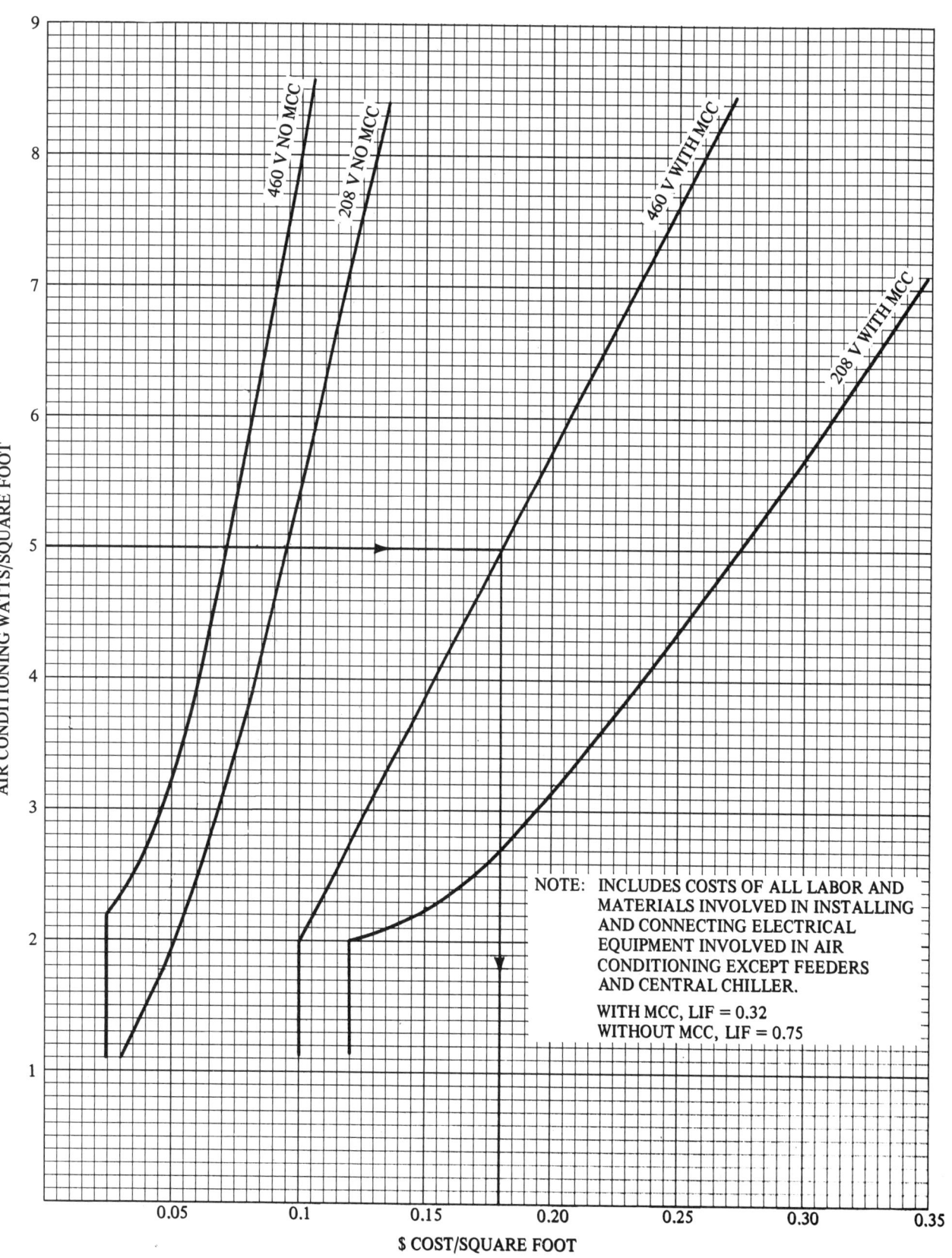

NOTE: INCLUDES COSTS OF ALL LABOR AND MATERIALS INVOLVED IN INSTALLING AND CONNECTING ELECTRICAL EQUIPMENT INVOLVED IN AIR CONDITIONING EXCEPT FEEDERS AND CENTRAL CHILLER.
WITH MCC, LIF = 0.32
WITHOUT MCC, LIF = 0.75

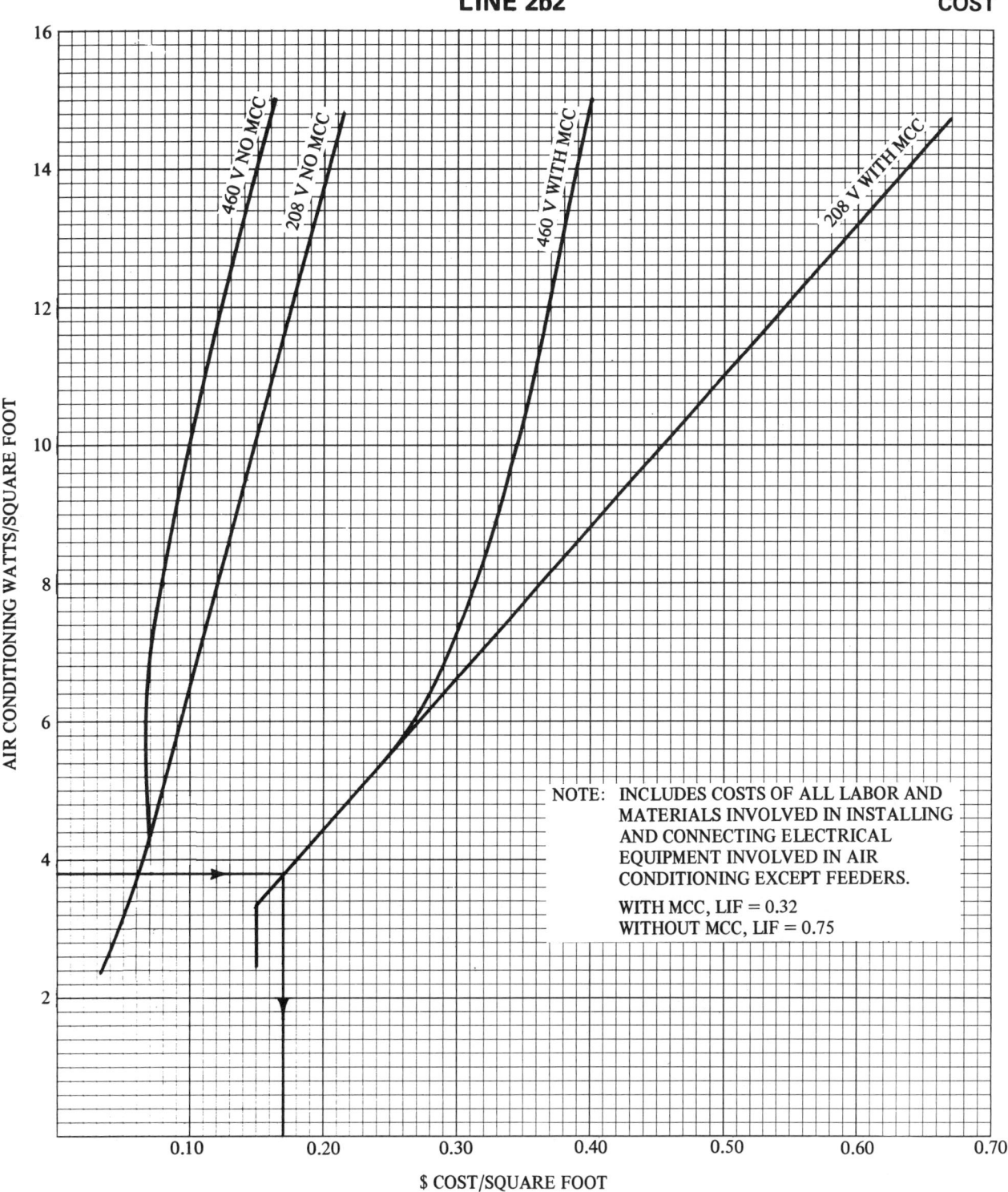

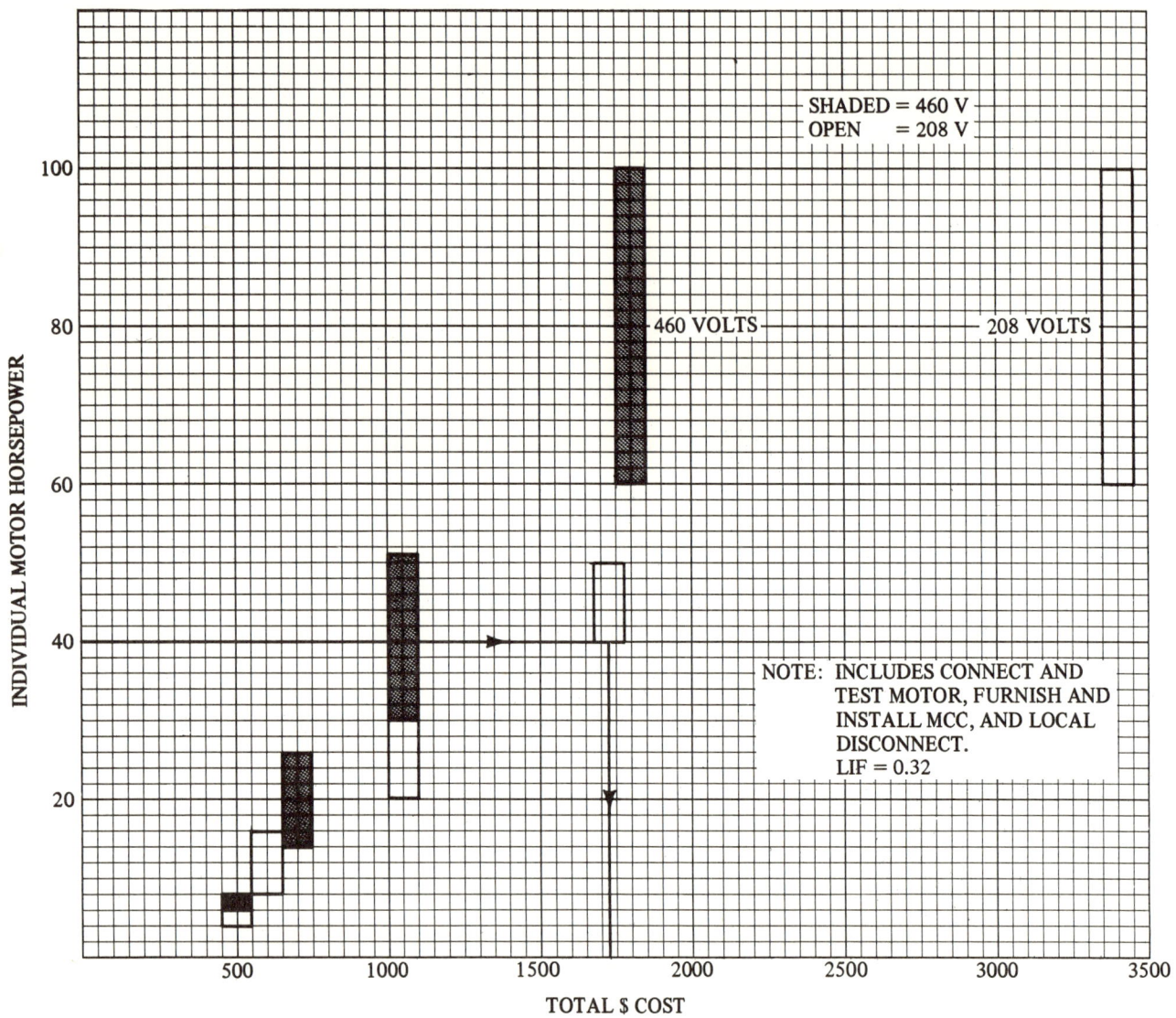

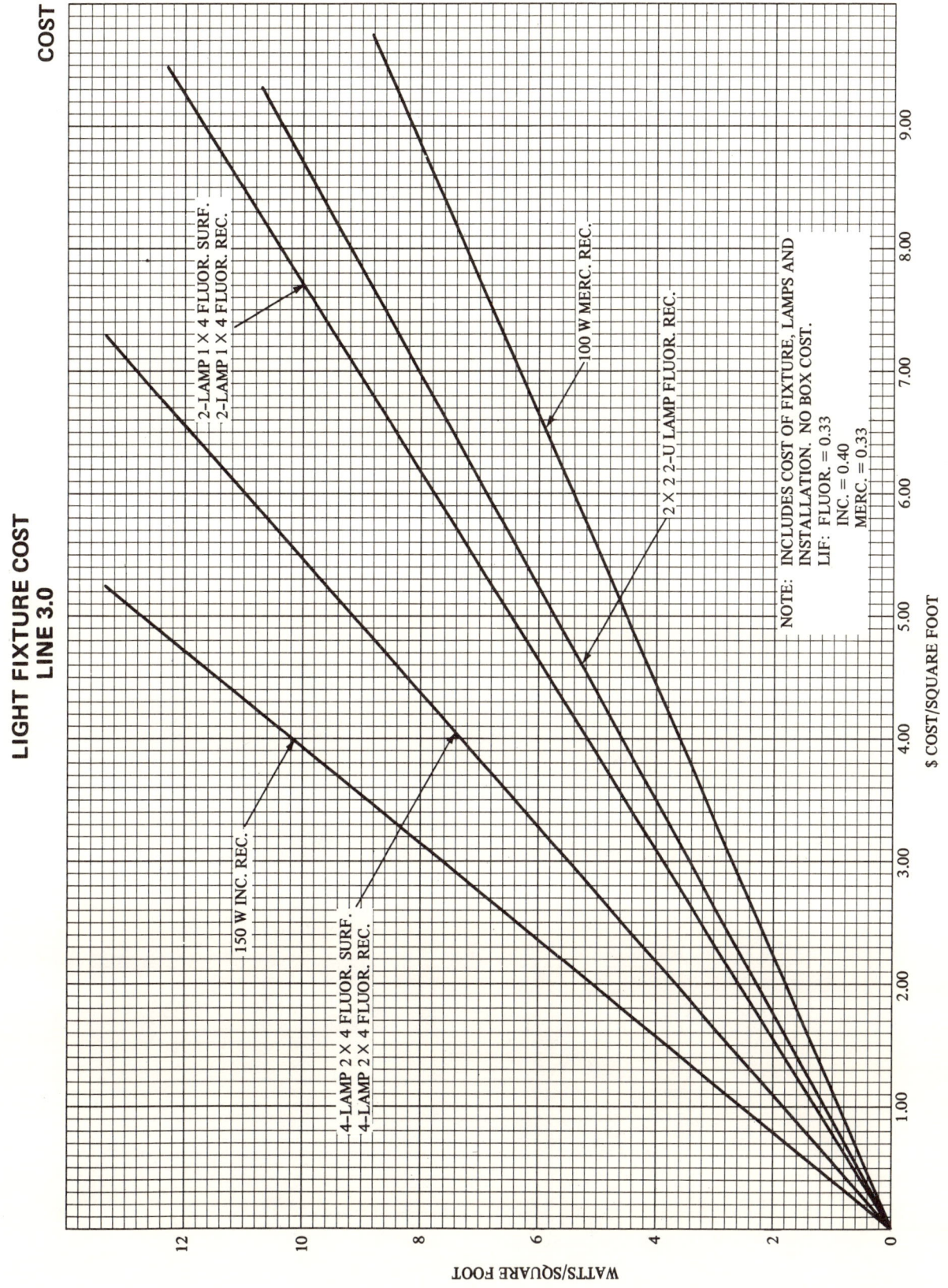

CHARTS 151

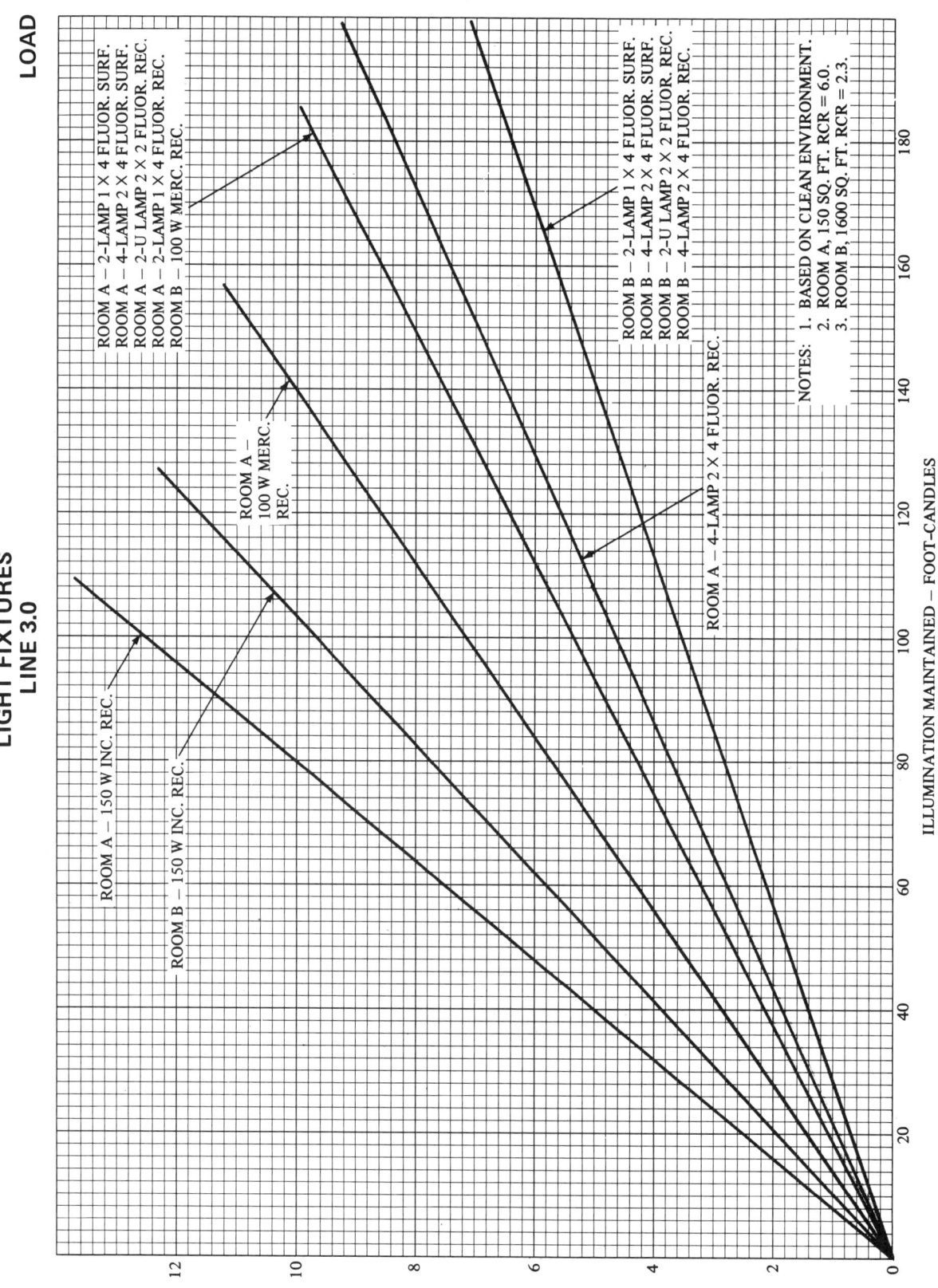

152 ESTIMATING AND COST CONTROL IN ELECTRICAL CONSTRUCTION DESIGN

6.
Sample Estimates

SUMMARY

In this chapter, a typical sample estimate, using the charts in the preceding Chapter 5, is reviewed. Where necessary, so-called subroutines have been illustrated. These are used to summarise certain information obtained from several of the charts and/or other sources. Whether or not these subroutines are used, any method should be consistent and systematic. Assumptions and notes should be explicit and all-inclusive.

ESTIMATE FORM

In the preceding chapters we have discussed all the components of an estimate. All the job-related factors, as well as the periodic estimate types, have been enumerated and examined for their cost effects. This information, as well as that obtained from the charts for the line items, is utilized to assemble an estimate. It should be pointed out again that regardless of which periodic estimate is being prepared and regardless of the amount of detail required at any of the phases, the format used should be identical for the life of the design phase. Much confusion and grief is generated by cost items slipping around from place to place as estimates are revised and updated. Projects may be halted or interrupted for major redesign at any phase. On such a job, two or three schematic estimates may be required, in which case it should be quite clear which set of documents is being estimated. All estimates should bear a set of definitions, a list of documents used, dates, and a clear delineation of the assumption made.

The general outline into which all the components fit is illustrated in Fig. 6-1. All cost items are assembled, properly factored, and collected to form the subtotal direct cost. Then the various markups and other operations are applied as illustrated in the electrical cost summary, (Fig. 6-2, in the next section).

Preparing an estimate is a process of examining the basic electrical building diagram (Fig. 3-14) for items to be included in the basic building; checking for the inclusion of other systems; selecting the proper charts; and then assembling the material on a line-by-line basis. Note that items numbered in the electrical diagram are numbered identically on the estimate reporting form. These items wil also have

ESTIMATE OUTLINE

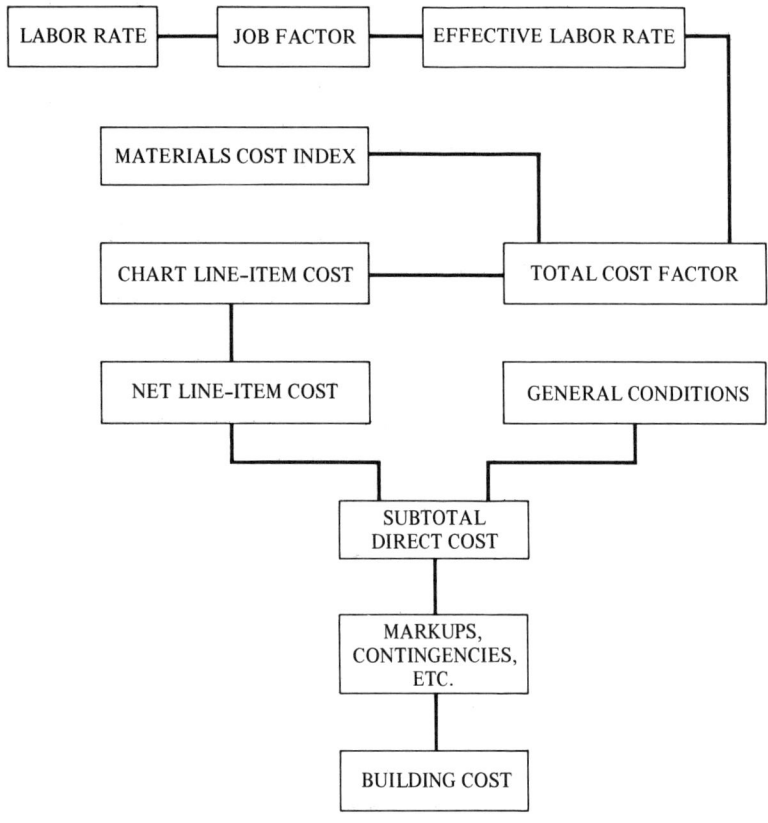

Fig. 6-1.

the same identifying number and title on the cost and load charts and in the line-item definitions in Chapter 4. For example, item 1b3, listed in the reporting form as "main distribution panel board," appears on the electrical diagram with the same designation, and is also described in Section 1b of Chapter 4, "Main Distribution Equipment, Secondary Service, and Metering." In these paragraphs there are data indicating the extent of the work and the electrical loads associated with each line item.

In order to assemble the electrical costs properly, data are provided in the form of load charts, which aid in first assembling electrical load data. We proceed from the electrical load to the sizing of the electrical equipment, and then to the pricing of that equipment. Knowing the load permits the equipment to be sized and therefore priced on a sound basis.

Since the electrical loads accumulate upstream toward the electrical service, load charts should be used in that order. Collecting motor loads and watts per square foot for lighting and receptacles does, by and large, give the total electrical load. Other point loads should be added when they become identified. Conversion charts allow these kilowatt loads to be converted to amperes, from which main service

ELECTRICAL COST SUMMARY

A.B.C. P.C.
Consulting Engineers

Estimate Type _____   Date of Estimate _____

Project _____

Location _____

Building Type _____   Gross Area _____

Building Description _____

Bid Date _____

Start Construction _____   Complete Construction _____

a. Direct Electrical Costs—Building		
b. General Conditions		
c. Subtotal Direct Cost		
d. Building Factor		
e. Design Contingency		
f. Building-Cost Subtotal		
g. Contractor Overhead and Profit		
h. Subtotal		
i. Escalation to		
j. Construction Contingency		
k. Building Cost Total		
l. Direct Cost—Site Work		
m. Design Contingency		
n. Contractor Overhead and Profit		
o. Subtotal—Site Work		
p. Escalation to		
q. Construction Contingency		
r. Site Work Total		
s. Construction Cost		

Fig. 6-2.

equipment and feeders may be sized and priced. Costs are charted for the more common voltages.

Some of the line items of cost require effort in addition to the direct reading of a related chart. In these cases, subroutines are illustrated showing the use of charted information to develop line-item electrical load and costs.

It must be borne in mind that the data on the charts are generalized. There is no substitute for hard takeoff and quotation estimating. As this information becomes available in the latter stages of the design process, it should definitely be used. Charts will find their greatest utility in the earlier design stages, when quantity takeoffs cannot be made.

> a. Direct Building Costs: including geographical adjustment.
>
> b. General Conditions: a percentage of line a.
>
> c. Subtotal Direct Cost: line a + line b.
>
> d. Building Factor.
>
> e. Design Contingency: a percentage of line c.
>
> f. Building-Cost Subtotal: line c + line e.
>
> g. Contractor Overhead and Profit: a percentage of line f.
>
> h. Subtotal: line f + line g.
>
> i. Escalation to Selected Date: a percentage of line h.
>
> j. Construction Contingency: a percentage of (line h + line i).
>
> k. Building-Cost Total: line h + line i + line j.
>
> l. Site-Work Cost.
>
> m. Design Contingency: a percentage of line l.
>
> n. Contractor Overhead and Profit: a percentage of line l.
>
> o. Site Subtotal: line l + line m + line n.
>
> p. Escalation to Date: a percentage of line o.
>
> q. Construction Contingency: a percentage of (line o + line p).
>
> r. Site-Work Total: line o + line p + line q.
>
> s. Construction Cost: line k + line r.

Fig. 6-3.

We have included a sample estimate for purposes of illustrating techniques described throughout this book. This estimate is for the program phase and makes heavier use of the charts than later estimates might. Nevertheless, the same principles and manner of assembling load and cost data can be used.

COST SUMMARY

In order to arrive at a bottom-line estimate, factors developed for the components outlined earlier in this chapter are assembled in a convenient form, such as that illustrated in Fig. 6-2 and explained in Fig. 6-3. Line *a* (direct costs) represents either a one-line summary of all the direct-cost components or a full list of these components. In either case, it is essential that all of the several multipliers be clearly listed and that the multiplier or percentage used be indicated along with the line title. Some of these factors may be changed in light of subsequent knowledge or upon direction of the client, and so should be clearly defined and separable from the body of the estimate. These same instructions apply to any of the periodic estimates. All components and assumptions should be very specific and visible. It is advisable to have a key to the assumptions and definitions attached to any estimate. This key might also include a negative list of items specifically deleted from the estimate for one reason or another. The purpose of this is to put the owner on guard for deliberate omissions of items of work or job factors that may have appeared in other projects of similar nature. In early estimates, items involving design decisions not part of the original program should be described where they form a basis for parts of the job pricing. All this serves to avoid negative answers to telephone calls that begin, "Did you include the cost of...," or "What about the...."

After summarizing the costs, it might be convenient to note the combined markup multiplier. This is, of course, the building-cost total divided by the building-cost subtotal, or line k divided by line f in Fig. 6-2. On occasion, whole items might be broken out of the estimate for discussion, and their total cost as part of the project would be their direct cost multiplied by the markup factor as determined above. This might be used for item-by-item comparison, for example, with a contractor who has included markups in each package of work.

In the case of Fig. 6-2, the electrical cost summary takes the form of a summary used for separate prime conract work. Visible here are all the markups the electrical prime contractor will apply. Determination of total project cost will then follow this general form, as an example:

Building
   General construction   1,000,000
   HVAC   200,000
   Plumbing   100,000
   Electrical   200,000
   Subtotal building cost   1,500,000

Site
   General construction site work   100,000
   HVAC site work   5,000
   Plumbing site work   40,000
   Electrical site work   20,000
   Subtotal site cost   165,000

        Total cost = $1,665,000
        A/E fee = 116,500
        Construction manager fee = 41,625
        Project cost = $1,823,125

Note that building work and site work are handled separately. This permits the convenient evaluation of cost per square foot for the building and permits separate handling for any of the cost markups where these may differ for site and building. In the event that work is performed under a general contractor or single prime contract, items of work may shift between contracts. Markups for components of the work may change. The estimator is obliged to anticipage all these costs and to demand clear delineation of contrasts by the designer.

SAMPLE ESTIMATE: PROGRAM PHASE

Better Health Nursing Home,
Baltimore, Maryland

The following are excerpts from the design program:
1. Total gross floor area: 200,000 square feet.
2. 100% air conditioned.
3. Lighting levels average 50 foot-candles.
4. Public telephone.
5. Master television antenna system.
6. Fire-alarm system using smoke detectors.
7. In-house food preparation, 150 kilowatt load.
8. Power will be purchased from the utility company and distributed at three-phase, four-wire, 480/277 volts.
9. Building will be 10 stories high.
10. Work is to be bid-packaged to provide wide participation of subbidders and will be run by a construction manager.
11. Half the floors will be typical.
12. Construction period will be two years, with completion scheduled for June 1980.
13. Three elevators at 20 horsepower each.
14. Site-work allowance is $100,000.

Assumptions

1. *Basic Labor Rate.* $10.71/hour.
   Use Fig. 3-1 to determine the job factor.
   Type of building: 2%
   Coordination: 2%
   8 floors above grade: 8%
                         ___
                         12%
   Effective wage rate = 10.71 × 1.12 = $12.00/hour
   Materials Cost Index = 0.98. Use 1.0.

2. *Line 3.0.* Assume one-half incandescent fixtures and one-half 2′ × 4′, 4-lamp fluorescent fixtures.

3. *Line 2c.* Assume motor control centers.
   Miscellaneous motors = 20% of air-conditioning motors.

4; *Line 2b1.* Assume small motor load is part of outlet load.

5. *Line 1e.* Assume emergency power requirements equal 10% of total power, or 200 KW.

6. *Line 1c4.* Power panels are nonexistent, except motor control centers already included.

7. *Line 1c1.* Assume the secondary service run from the utility-company vault will be 75 feet of 480-volt, 3-phase, 4-wire cable and conduit. Use 500 MCM sets. For 2500-amp present load use 7 sets.
7 × 380 = 2660.
Chart 1d1, 1d2: CLF = $23.00.
Total cost = 7 × $23.00 × 75 = $12.075.
Assume copper wire and galvanized conduit.

8. *Line 4a1.* Assume 200 telephones at $100 per instrument for conduit and box system.

9. *Line 4a2.* Assume there will be two small sound systems for a total of $8,000.

10. *Line 4a3.* Assume there will be 200 television outlets in the master antenna system.

11. *Line 4b1.* Assume a smoke-detection and manual fire-alarm station.

Procedure

1. Collect load data from 3.0 toward 1.0 as cost of items is noted on cost summary sheet.
2. Use load subroutine L1 to obtain motor loads (line 2b2).
3. Assume small-motor line 2b1 to be included with outlet requirements.
4. Use cost subroutine C1 to determine branch-circuit costs.
5. Use charts to find device and lighting outlet box costs.
6. Summarize power and light load to give total power requirements. Use the stated percentage to select an emergency generator size and unit cost.
7. Use previously determined receptacle electric load to determine dry-type transformer costs.
8. Use number of branch circuits obtained from the charts to determine the lighting and receptacle panel cost for lines 1c1 and 1c2. Observe the proper voltage mix of the devices.
9. Use the KW–amperes conversion chart (Chapter 4) with KW load previously determined to find service equipment ampere rating.

Subroutine L1

Motor Loads	Data Source	Quantity	Watts/SF	Area	Watts total load
1. A/C Load	Appendix	400 SF/ton	3.2	100%	640,000
2. Other Motors	Assume	20% A/C	0.64	100%	128,000
3. TOTAL	Chart 2b2		3.84	200,000	768,000

Subroutine C1

Branch Circuit (2a4)	Data Source	Quantity	C/SF	Area	Total cost
1. Light Switch	Chart 2a4		0.16	100%	32.000
2. Receptacle	Chart 2a4		0.36	100%	72.000
3. Lighting	Chart 2a4		0.34	100%	68.000
4. TOTAL			0.86	200,000	172.000

LOAD SUMMARY SHEET
Sample Estimate—Nursing Home, Baltimore

		Data Source	Unit	Unit Load	Quantity or Applicable Area	Total Load-Watts
1.0	Service & Distribution					
1a1	Primary wire and conduit					
1a2	Primary service switch					
1a3	Power transformer					
1b1	Secondary service wire and conduit					
	Basic Building					
1b2	Secondary service & metering					
1b3	Main distribution equipment					
1c1	Lighting panel 480/277 v					
1c2	Light & receptacle panel 208/120 v					
1c3	Dry type transformer					
1c4	Power panel					
1d1	Motor feeders					
1d2	Lighting feeders					
1e	Emergency power					
2.0	Power requirements					
2a1	Lighting outlet box					
2a2	Receptacle	Chart 2a4	W/SF	1.05	200000	210000
2a3	Light switch	—				
2a4	Branch circuits	—				
2b1	Small motor & control	—				
2b2	Large motor & control	Subroutine #L1				768000
2c	Motor control center	—				
2d	Kitchen	Program				150000
3.0	Illumination					
	150W Incandescent	Chart 3.0	W/SF	3.8	100000 SF	380000
	2 X 4 Fluorescent	Chart 3.0	W/SF	4.0	100000 SF	400000
	Total Load 2 + 3					1908000
	Watts/Sq Ft					9.5

COST SUMMARY SHEET NO. 1
Sample Estimate—Nursing Home, Baltimore

	Data Source	Unit	Unit Cost	Quantity or Area	Subtotal	L.I.F.	M.C.I.	12.00 Labor T.C.F.	Total Cost	
1.0 Service & Distribution										
1a1 Primary wired conduit										
1a2 Primary switch & fuse										
1a3 Power transformer										
1b1 Secondary service wired										
Conduit	Chart 1d1; 1d2	CLF	23.00	525	12075	0.78	1.0	0.68	8211	
Basic Building										
1b2 Secondary service & metering	Chart 1b2	L.S.		200000	12000	0.14	1.0	0.94	11280	
1b3 Main distribution equipment	Chart 1b3	L.S.		200000	8600	0.31	1.0	0.88	7568	
1c1 Lighting panel 480/277 v	Chart 1c1; 1c2	CSF	0.05	200000	10000	0.41	1.0	0.84	8400	
1c2 Light & receptacle panel 208/120 v	Chart 1c1; 1c2	CSF	0.11	200000	22000	0.63	1.0	0.72	15840	
1c3 Dry type transformer	Chart 1c3	KCF	37.00	200000	7400	0.37	1.0	0.86	6364	
1c4 Power panel										
1d1 Motor feeders	Chart 1d1	CSF	0.73	200000	146000	0.80	1.0	0.68	99260	
1d2 Lighting feeders	Chart 1d2									
1e Emergency power	Chart 1e	CSF	0.11	200000	22000	0.05	1.0	1.00	22000	
2.0 Power requirements										
2a1 Lighting outlet box	Chart 2a1	CSF	0.11	200000	22000	0.95	1.0	0.62	13640	
2a2 Receptacle	Chart 2a2	CSF	0.13	200000	26000	0.84	1.0	0.66	17160	
2a3 Light switch	Chart 2a3	CSF								
2a4 Branch circuits	Subroutine #C-1	CSF	0.86	200000	172000	0.78	1.0	0.68	116960	
2b1 Small motor & control	—	—								
2b2 Large motor & control	Chart 2b2	CSF	0.17	200000	34000	0.32	1.0	0.88	29920	
2c Motor control center	—	—								
2d Kitchen	Program	L.S.			8000	0.32	1.0	0.88	7040	
3.0 Illumination										
150W Incandescent	Chart 3.0	CSF	1.50	100000	150000	0.40	1.0	0.84	126000	
2 × 4 Fluorescent	Chart 3.0	CSF	2.40	100000	240000	0.33	1.0	0.88	211200	
					898000				700863	

COST SUMMARY SHEET NO. 2
Sample Estimate—Nursing Home, Baltimore

	Data Source	Unit	Unit Cost	Quantity or Area	Subtotal	L.I.F.	M.C.I.	(12.00 Labor) T.C.F.	Total Cost
4.0 Other Systems									
4a Communications									
4a1 Telephone	Assume	L.S.		200	20000	0.33	1.0	0.86	17200
4a2 Public address	Assume	L.S.		2	8000	0.40	1.0	0.84	6720
4a3 Television antenna system	Chap. 4	BACH	75.00	200	15000	0.37	1.0	0.86	12900
4a4 Clock System	—								
4a5 Nurse Call System	—								
4b Security									
4b1 Fire Alarm	Quote			200000	100000	0.35	1.0	0.86	86000
4b2 CCTV	—								
4c Miscellaneous Systems									
4c1 Lightning Protection	—								
4c2 Temporary Light and Power	Chap. 4	CSF	0.13	200000	26000	0.40	1.0	0.84	21840
								Page 2 Total	144640
								Page 1 Total	700863
								Grand Total	845503

164 ESTIMATING AND COST CONTROL IN ELECTRICAL CONSTRUCTION DESIGN

Appendix

SUMMARY

Additional useful general information is included here. There are charts and schedules to aid in estimating pump motor and air conditioning horsepower requirements. Other forms are assembled so that they can be duplicated by the user for his convenience. There is also a brief bibliography.

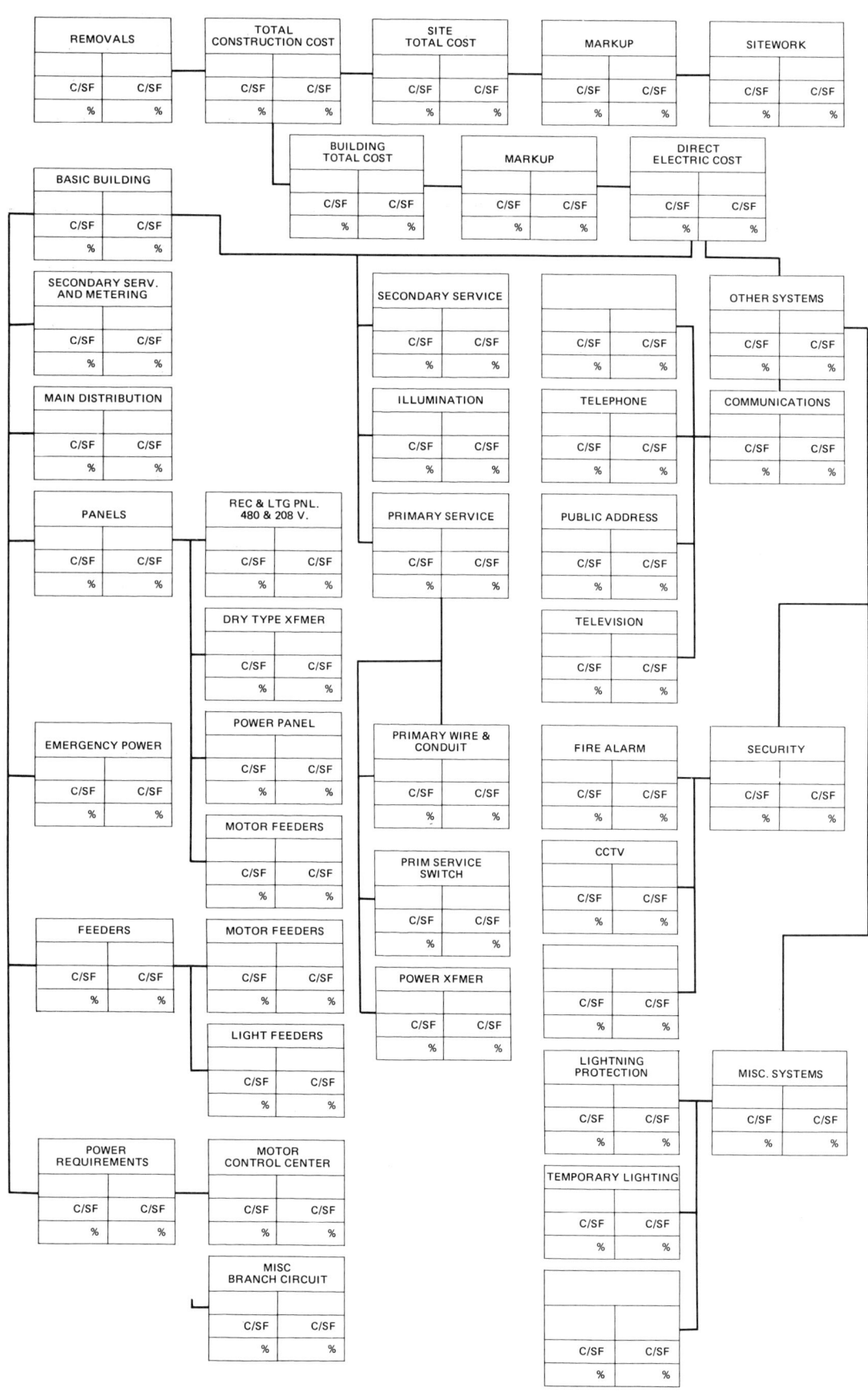

NOTE: COST BLOCK FIGURES ARE WITHOUT OVERHEAD AND PROFIT

ELECTRICAL COST MODEL—Fig. CM-1

Dodge Digest of Building Costs and Specifications

PAGE: D 201   ISSUE DATE: 9/76   EDITION #: 22

BUILDING TYPE: COMMERCIAL GARAGES

#	PROJECT & LOCATION	SHAPE, DIMENSIONS & STRUCTURE	EXTERIOR WALL	ROOF	FLRNG FLR COV	INTERIOR WALL	# STORIES & UNITS	BID DATE	STRUCT	PLUMB	HVAC	ELECT.	MISC.	TOTAL	CUBIC FEET	SQUARE FEET
1	Service Sta / Upland CA / 900	Sq; 24'x62' wd truss	mas	tile	conc	wd stud drywl	1 sty-no bsmt	05/75	21	2	3	11	1	38	TOTAL CF 11,904 / $/CUBE 3.20	TOTAL SF 1,488 / $/SQUARE 25.59
2	Carwash Gas Sta / Jefferson Co. MO / 631	Rect; 30'x50' wl brg	blk	precast conc	slab on grd VAT	drywl & blk	1 sty-no bsmt	09/75	34 / 37,000 BTU htg - 3 ton AC.	10	4	8	58	114	TOTAL CF 18,000 / $/CUBE 6.31	TOTAL SF 1,500 / $/SQUARE 75.67
3	Maint. Bldg & Pkg Lot / Eau Claire WI / 544	Rect; 35'x56' single span conc dbl tee on mas brg wls	conc blk	b u	exp conc	conc blk	1 sty-no bsmt 157 pkg spaces	11/75	57 / 600 M/BTU elec basebd 30 kw htg.	5	5	5	-----	72	TOTAL CF 24,750 / $/CUBE 2.89	TOTAL SF 1,980 / $/SQUARE 36.12
4	Garage / Somerset Co. NJ / 086	Rect; 40'x50' wd truss	blk	------	conc	pt cer t	1 sty-no bsmt	09/75	39 / 320 M/BTU oil fired f a htg.	5	4	5	-----	53	TOTAL CF 33,000 / $/CUBE 1.59	TOTAL SF 2,000 / $/SQUARE 26.31
5	Maint. Gar. / Berks Co. PA / 181	Rect; 46'x50' stl rf & brg wls	12" conc blk ptd	stl dk 2" ins b u	conc slab on grd	unfin	1 sty-no bsmt 1 unit	10/74	47 / 100 M/BTU oil fired h a htg.	16	8	6	-----	78	TOTAL CF 41,400 / $/CUBE 1.88	TOTAL SF 2,300 / $/SQUARE 33.80
6	Maint. Shop / Decatur MS / 392	Rect; 63'x38' ld brg wls	brk	bar jsts & mtl dk	conc slab on grd exp	CMU ptd	1 sty-no bsmt 2 bays	06/75	48 / Oil fired air @ serv areas & elec rad @ ofcs htg.	4	4	6	-----	63	TOTAL CF 34,162 / $/CUBE 1.84	TOTAL SF 2,394 / $/SQUARE 26.27
7	Service Sta / Prince George's Co. MD / 212	Irreg; 76'x48' blk & fr wls	tex ply wd	asph shgls on wd truss	slab on grd	ptd ply wd	1 sty-no bsmt	10/74	48 / 185 M/BTU elec htg - no AC.	24	5	10	3	91	TOTAL CF 27,360 / $/CUBE 3.33	TOTAL SF 2,912 / $/SQUARE 31.25
8	Service Sta. / Bolingbrook IL / 606	Rect; brg wl	f brk	bar jst	conc slab on grd	conc blk	1 sty-no bsmt	06/74	30	21	10	10	42	115	TOTAL CF ------ / $/CUBE ------	TOTAL SF 3,000 / $/SQUARE 38.22
9	Storage & Maint. / Hopewell NY / 132	Rect; 102'x30' stl jsts wl brg	conc blk	mtl dk on stl jsts	conc slab on fill ptd	conc blk	1 sty-no bsmt	09/74	60	5	-----	5	-----	69	TOTAL CF 42,840 / $/CUBE 1.61	TOTAL SF 3,060 / $/SQUARE 22.60
10	Garage (Bus) / Lawrenceburg KY / 402	Rect; 64'x50' rigid stl fr & purlins	mtl wl pnl	mtl pnls	conc slab on grd	mtl liner conc blk	1 sty-no bsmt 1 unit	07/75	39 / 160 M/BTU gas fired h w htrs & fin tube rad htg - no AC.	12	4	6	6	66	TOTAL CF 64,000 / $/CUBE 1.04	TOTAL SF 3,200 / $/SQUARE 20.75

167

DODGE BUILDING COST AND SPECIFICATION DIGEST

PAGE E 106 — HOSPITALS — ISSUE DATE 3/71

NO.	LOCATION	BID DATE	STORIES	STRUCTURE	EXT.	ROOF	FLOOR	INT. WALL	TOTAL SQUARE	TOTAL CUBE	STRUCT.	PLUMBING	H. & V.	A/C	ELECT.	MISC.	TOTAL	$/SQ.	$/CUBE
51	Hospital Newport, Rhode Island	7/66	8 & bas.	rein. conc. strucl. syst.	brk. & l.s.; conc. blk. bk-up	rein. conc. b.u.	cer. & vinyl asb. tile carpet	vinyl fab. cer.tile pt. wall	168,873	1,870,000	3,314	450	429	634	980	*452	6,259	37.06	3.35
	Rectangle - 85' x 165' -- Basmt. - dietary, laun. central sterile supply, inhal. therapy & pharmacy; 1st flr. obstet.-gynecol. nursing unit; 2nd flr.-surg. suite & postoper. recov. rm.; 3rd flr.mech.equip.etc. Oil fired stm. boilers w/h.w. conv. htg. syst.,fan coil units air cond. syst. *Built-in equip.,elevs.																		
52	Children's Hospital Kansas City, Missouri	11/67	5 & bas.	rein. conc. strucl. syst.	precast conc. brk.; mas.blk. bk-up	conc. flat slab	vinyl asb.tile carpet spec. coat'gs.	gyp.bd. mtl.studs ptd. & coatings	172,736	2,559,433	3,045	--	--	1,482	487	--	5,013	29.02	.20
	Square flrs. 1 & 2, Rectangular flrs. 3,4 & 5 - 243' x 224' -- 100 beds, labs., radiology, surgery, phys. ther., occup. ther., out-pat. facils., din. rm., kitch. & admin. All air syst., low vel. dual ducts admin. & lower flrs., low vel. sgl. duct reheat syst. nursing units. 22,000,00 BTU htg. & 834 tons air cond.																		
53	Hospital Lansing, Michigan	11/69	7 & bas.	stl. frm. strucl. syst.	brk.; l.w. blk. bk-up	conc. mtl. dk.	cer. & vinyl asb. tile	pt. & vinyl fab. on plast.	191,810	3,129,096	5,079	--	2,431	--	1,070	*377	8,958	46.70	2.85
	Rectangular - 140' x 355' -- 260 beds. Air induction units htg. & air cond. syst. in patient rms. *Elevs. & dumbw.																		
54	Hospital Savannah, Georgia	1/67	7 & part bas.	rein. conc. strucl. syst.	brk. & cast stone; conc.blk. bk-up	conc.	cer. quarry resil. tile carpet	strucl. glz.tile cer. plast. wl.cov'g.	199,982	2,569,699	916	522	--	682	484	*412	3,016	15.08	1.75
	Square (1st flr.) & Cross (upper flrs.) - 231' x 271' -- 200 beds from 2nd thru 5th flrs.; pneu. tube, elevs. Stm. htg. syst., 20,700 lbs. hr. 690 tons air cond. *Built-in equip.																		
55	Hospital Amarillo, Texas	3/66	6 & part bas.	rein. conc. strucl. syst.	exp.agg. conc. & brk.; conc.blk. bk-up	b.u. pitch pan jst. bms. girders	vinyl asb.tile terr. & quarry & cer.tile	plast. cer.tile vinyl wl. covr'g. mbl.glz.T	200,815	2,689,770							4,700	23.40	1.75
	Cross - 425' x 355' (1 & 2), 250' x 245' (3-4-5-6); 240 beds built to expand to 480 beds by addl. flrs., central facils., labs., cafet., oper. rms. now accom. 480; stand-by gen.,cbnt. work, pneu. tube & convey.incl. Gas-fired high press. stm. htg. syst.; 600 tons tot. air cond. syst.																		
56	Hospital Philadelphia, Pennsylvania	8/67	8 & bas.	rein. conc. strucl. syst.	l.s. & brk.; conc. blk. bk-up	conc. & b.u.	terr. & vinyl asb. tile	cer.tile pt. & vinyl fab. on plast.	222,815	2,660,970	5,100	711	--	946	947	--	7,704	35.00	2.90
	2 circ. towers joined by link - 266' x 110'. 4 pipe hot & chilled w. syst. & air handl'g. units, central plant stm. to hot w. htg. syst.; 810 tons air cond., htg. 16,000,000 BTU.																		
57	State School & Hospital Philadelphia, Pennsylvania	2/70	1 & part bas.	wall br'g. strucl. syst.	brk.; conc. blk. bk-up	insul. conc. bonded	vinyl asb. tile	ptd. conc. blk.	257,881	4,540,150	6,932	1,028	--	1,720	1,570	--	11,250	43.60	2.49
	18 bldgs. var. shapes & sizes - Sitework.																		
58	Hospital Phoenix, Arizona	6/67	7 & bas.	rein. conc. strucl. syst.	brk. & text. conc.blk. bk-up	conc. slab b. u.	colored conc. Torginol quarry vinyl T.	cer.tile plast. vinyl glz.	419,680	5,063,824	5,536	982	--	1,706	1,148	*740	10,112	24.09	1.99
	"T" & Cross shaped - 320' x 440'. Pneu. tube syst., vacuum laun. syst., waste handl'g. syst., large out-patient dept. & clinic; 489 beds. Stm. 64,000 # cap., high press. htg. syst., 2 boilers; multi-zone air handl'g. units, chilled w. syst., 2,000 tons, 2 gas driven chillers w/exh. h. recov. driv., 600 T. *Built-in equip.																		
59	Hospital Black River Falls, Wisconsin	6/66	3 & part bas.	rein. conc. strucl. syst.	brk.; blk. furred plast. bk-up	conc. pan	cer.tile carpet vinyl asb.	plast. vinyl fab. cer.tile	63,887	793,110	773	129	--	316	162	*40	1,420	22.50	1.79
	"T" shaped - 235' & 42' & 180' x 73' -- 76 beds (des. for fut. 36 bed flr.), 2 surgery rms., kitch., 2 rm. X-ray suite, lab, emerg., phys. therapy, stor., 2 elevs., all supporting facils. des. for 112 beds. High press. stm., two-270 HP boilers htg. syst.; 188 tons air cond. syst. *Built-in equip.																		

BOARD OF EDUCATION. CITY OF NEW YORK
BUREAU OF CONSTRUCTION

SCHEDULE OF ITEMS AND COSTS

School _____ Borough _____

Contractor _____ Architect _____

The total cost of each of the following items and sub-divisions thereof, must be inserted in the columns so captioned opposite same. "Costs" shall represent the true value of the work installed and in addition contain in each item the pro rate share of the contractor's overhead and profit. The space opposite items not included in the contract are to be left blank. Any items of work included in the contract, but not listed, must be inserted in the numbered blank spaces provided.

This Schedule must be dated and signed by the contractor or an officer of the firm in the space provided and returned to the Director (Division of Design and Construction) as provided for in the contract.

ITEM NO.	ITEM	COSTS NEW BLDG. OR ADDITION	COSTS EXIST. BLDG. MODERN	TOTAL COSTS	%
1A	Insurance				
B	Bond Premium				
2	Temporary Light and Power Installed				
3A	Conduits—Basement Floor				
B	First Floor Slab				
C	Second Floor Slab				
D	Third Floor Slab				
E	Fourth Floor Slab				
F	Roof Slab				
4A	Nippling—Basement				
B	First Floor				
C	Second Floor				
D	Third Floor				
E	Fourth Floor				
F	Roof Conduit				
5A	Conduits between Main Switchboard and Panel Box Locations including back boxes				
6	Conduits between interconnecting Box Locations				
7	Conduits between Sound Control Cabinet and L.S. Box Locations				
8	Conduits to TV Rack, between TV outlets and TV interconnecting box				
9	Conduits between Clocks and Terminal Boxes				
10	Conduits and Boxes in Hung Ceilings				
11A	Conduits—Underground Electric Service				
B	Conductors Service Feeders				
12A	Fire Alarm System—Exterior Conduits and New Manholes Installed				
B	Exterior Cable installed				
13	Telephone Conduits Installed, Interior System				
14A	Wiring for Light and Power—Main Feeders				
B	Branch Circuit Wiring				
15	Wiring for Sound				
16	Wiring for Clock System				
17	Wiring for Fire Signal System—Interior, and Smoke or Heat Detector System				
18	Switches, Receptacles and Plates				
19	Fire Signal and Detection System Equipment, and Control Boards				
20	Interior Telephone Switchboard, Equipment, and Back Boxes				
21A	Panel Boards—Light				
B	Power—Shops, Fan Rooms, Boiler Room, etc.				
22	Stage Switchboards and Dimmer Board Installed Complete				
23	Motor Starters and Controls, Motor Control Centers				
24	Main Switchboard, Sub-Distribution Center installed and Wired Complete				
25	Electric Service and Metering Equipment				
26	Loud Speakers and Clocks				
27	Sound Control Cabinet Installed and Wired Complete				
28	Television System Rack, TV Outlets, Installed and Wired Complete				
29	Program System Equipment Installed, Wired and Tested Complete				
30	Time Recorder. Complete (Total number.......)				
31	Emergency Lighting System Equipment Installed and Wired Complete				
32A	Light Fixtures—Basement				
B	First Floor				
C	Second Floor				
D	Third Floor				
E	Fourth Floor				
F	Cafeteria				
G	Auditorium				
H	Gymnasium				
33	Lamps and Glassware				
34A	Stage Lighting—Border Lights				
B	Soot Lights				
C	Proscenium Lights				
35	Language Lab Equipment Installed and Wired Complete (......Rms)				
36	Sprinkler Alarm System, Complete				
37	Patching and Painting				
38	Misc. Electrical Equipment—Kilns, Grinders, Eraser Cleaners, Dental Equipment				
39	Removals				
40	Miscellaneous Items				
41	Testing, Adjustments and Certificates				
	TOTALS				

Contractor _____ Date _____
(Signature of Contractor or Officer of Firm)

Architect or Engineer _____ Date _____

B. of E. Chief of Electrical Design Section _____ Date _____

Typical Illumination foot-candle recommendations by Illuminating Engineering Society Handbook

Hospitals
    Anesthetizing and preparation room 30
    Autopsy and morgue
        Autopsy room 100
        Autopsy table................... 1000
        Museum....................... 50
        Morgue, general................ 20
    Central sterile supply
        General, work room............. 30
        Work tables 50
        Glove room 50
        Syringe room 150
        Needle sharpening 150
        Storage areas 30
        Issuing sterile supplies 50
    Corridor
        General in nursing areas—daytime..... 20
        General in nursing areas—night
            (rest period) 3
        Operating, delivery, recovery, and
            laboratory suites and service
            areas 30
    Cystoscopic room
        General...................... 100
        Cystoscopic table 2500
    Dental suite
        Operatory, general............... 70
        Instrument cabinet 150
        Dental entrance to oral cavity 1000
        Prosthetic laboratory bench......... 100
        Recovery room, general,........... 5
        Recovery room, local for observation .. 70
    (EEG) encephalographic suite
        Office (see Offices)
        Work room, general............. 30
        Work room, desk or table 100
        Examining room 30
        Preparation rooms, general 30
        Preparation rooms, local........... 50
        Storage, records, charts........... 30

Electromyographic suite
    Same as EEG but provisions for reducing
        level in preparation area to 1
Emergency operating room
    General..................... 100
    Local 2000
EKG, BMR, and specimen room
    General..................... 30
    Specimen table 50
    EKG machine.................. 50
Examination and treatment room
    General..................... 50
    Examining table 100
Exit stairways and landings, on floor 5
Doorways...................... 10
Administrative and lobby areas, day..... 50
Administrative and lobby areas, night.... 20
Chapel or quiet area, general......... 5
Chapel or quiet area, local for reading ... 30
Physical therapy.................. 20
Occupational therapy 30
Work table, course work 100
Work table, fine work 200
Recreation area 50
Dining area 30
Patient care unit (or room), general 20
Patient care room, reading 30
Nurse's station, general
    Day .... 50†
    Night 20
Nurse's desk, for charts and records..... 70†
Nurse's medicine cabinet .... 100†
Utility room, general................ 20
Utility room, work counter 50
Pharmacy area, general 30
Pharmacy compounding, and dispersing
    area 100
Janitor's closet................... 15
Toilet and bathing facilities 30
Barber and beautician areas 50

Equivalent Full-Load Hours of Operation Per Year

	Atlanta	Baltimore	Boston	Chicago	Dallas	Denver	Detroit	Los Angeles	Miami	Milwaukee	Minneapolis	New Orleans	Okla. City	Philadelphia	Phoenix	Portland Ore.	San Fran.	Saint Louis	Wash., D.C.	New York
Restaurants	1750	1620	1050	1250	2240	1050	1250	1150	2020	1050	1050	2020	2240	1480	2240	1050	450	2020	1820	1430
Drug Stores	1700	1580	1030	1220	2170	1030	1220	1120	1850	1030	1030	1950	2170	1440	2170	1030	400	1950	1580	1400
Cafeterias	1370	1270	825	990	1750	825	990	910	1580	825	825	1580	1750	1160	1750	825	350	1580	1270	1120
Jewelry Stores	1020	950	620	750	1300	620	750	700	1170	620	620	1170	1300	875	1300	620	250	1170	950	850
Barber Shops	1020	950	620	750	1300	620	750	700	1170	620	620	1170	1300	875	1300	620	250	1170	950	850
Night Clubs	1010	940	610	730	1280	610	730	675	1150	610	610	1150	1280	860	1280	610	240	1150	940	840
Theaters	650–1000	600–1000	400–650	500–800	850–1400	400–650	500–800	475–750	800–1300	400–650	400–650	800–1300	850–1400	550–920	850–1400	400–650	200–400	800–1300	600–1000	550–900
Dress Shops	940	870	565	675	1200	565	675	630	1080	565	565	1080	1200	800	1200	565	225	1080	870	780
Large Offices	915	850	550	660	1180	550	660	610	1060	550	550	1060	1180	775	1180	550	200	1060	850	750
Department Stores	850	790	515	650	1100	515	650	600	1000	515	515	1000	1100	725	1100	515	175	1000	790	700
Specialty Shops (5 & 10)	840	780	510	640	1080	510	640	590	975	510	510	975	1080	710	1080	510	175	975	780	700
Residences	810	750	490	600	1050	490	600	550	950	490	490	950	1050	690	1050	490	170	950	750	670
Shoe Stores	650	600	400	500	850	400	500	475	775	400	400	775	850	575	850	400	150	775	600	550
Beauty Shops	625	580	380	450	800	380	450	425	750	380	380	750	800	540	800	380	150	750	580	525
Small Offices	540	500	425	450	700	425	450	410	650	425	425	650	700	490	700	425	125	650	500	475
Recreation Spaces	520	480	450	400	675	450	400	380	650	450	450	650	675	470	675	450	125	650	480	450
Funeral Parlors	460	425	350	375	600	350	375	350	575	350	350	575	600	410	600	350	100	575	425	400

From Trane Air Conditioning Manual

Commercial Cooking Equipment

Appliance	Capacity	Overall Dim., Inches Width × Depth × Height	Miscellaneous Data (Dimensions in Inches)	Watts
Electric, Floor Mounted Type				
Broiler, no oven			23 wide × 25 deep grid	12,000
With oven			23 × 27 × 12 oven	18,000
Deep fat fryer	28 lb fat	20 × 38 × 36	14 wide × 15 deep kettle	12,000
	60 lb fat	24 × 36 × 36	20 wide × 20 deep kettle	18,000
Oven, baking, per sq ft of hearth			Compartment 8-in. high	500
Oven, roasting, per sq ft of hearth			Compartment 12-in. high	900
Range, heavy duty		36 × 36 × 36		15,000
Top section				6,000
Oven				
Range, medium duty		30 × 32 × 36		8,000
Top section				3,600
Oven				
Range, light duty		30 × 29 × 36		6,600
Top section				3,000
Oven				
Electric, Counter Type				
Coffee brewer				
Per burner				625
Per warmer				160
Automatic	240 cups per hr	27 × 21 × 22	4-burner + water htr.	5,000
Coffee urn	3 gal.			2,000
	5 gal.			3,000
	8 gal. twin			4,000
Deep fat fryer	14 lb fat	13 × 22 × 10		5,500
	21 lb fat	16 × 22 × 10		8,000
Dry food warmer, per sp ft of top				240
Egg boiler	2 cups	10 × 13 × 25		1,100
Griddle, frying per sq ft of top				2,700
Griddle-Grill		18 × 20 × 13	Grid. 200 sq in.	6,000
Hotplate		18 × 20 × 13	2 heating units	5,200
Roaster		18 × 20 × 13		1,650
Roll warmer		18 × 20 × 13		1,650
Toaster, continuous	360 slices/hr	15 × 15 × 28	2 slices wide	2,200
	720 slices/hr	20 × 15 × 28	4 slices wide	3,000
Toaster, pop-up	4 slice	12 × 11 × 9		2,540
Waffle iron		18 × 20 × 13	2 grids	1,650

ASHRAE Handbook of Fundamentals, 1972.

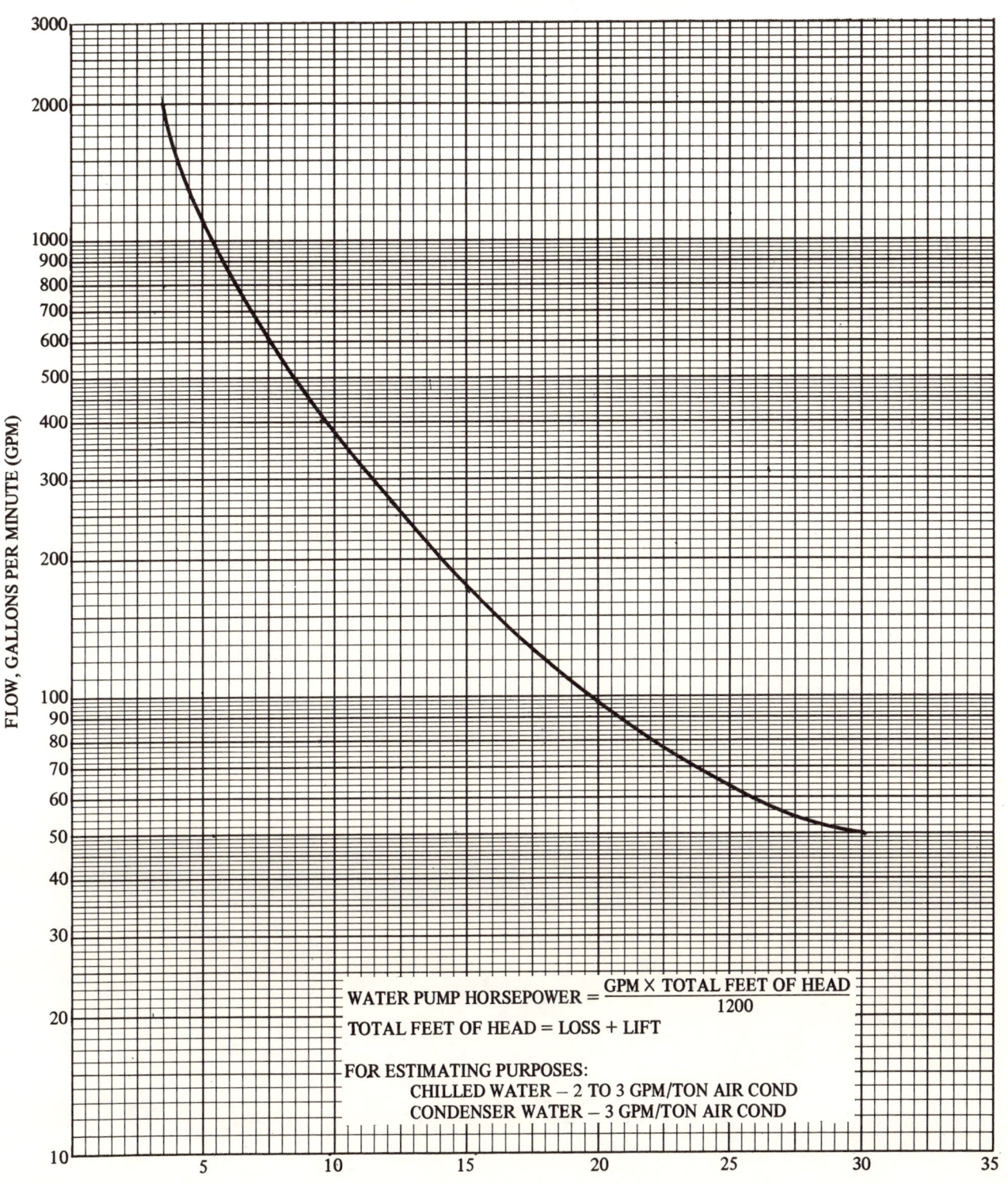

Estimating Air-Conditioning Motor Loads

Building type	Square feet of building per ton A/C	CFM air supply per square feet of building
Apartment	900–400	0.9–0.5
Museum	400–160	2.1–0.9
Bank	350–160	2.5–1.1
Department store	600–300	2.0–0.9
Office	400–200	2.2–0.7
Theater	150–100	15–30*
Hospital	350–200	2.0–1.2
Nursing home	650–350	1.5–0.7
Public school	400–300	1.8–1.2

*CFM per person.

Compressor horsepower per ton average for comfort cooling = 1 horsepower per ton of air conditioning.

Air Conditioning Fan Horsepower Requirements

$$\text{Fan HP} = \frac{\text{CFM} \times \text{Static Pressure (Inches of water)}}{4,000}$$

Low-pressure system—S.P. = less than 3″.
Medium-pressure system—S.P. = 3″–6″.
High-pressure system—S.P. = above 6″.

LIFE CYCLE COSTS

ITEM:

INITIAL COSTS	ORIGINAL	ALT. 1	ALT. 2	ALT. 3
Base Cost				
Interface Costs				
(a)				
(b)				
Other Initial Costs				
(a)				
(b)				
TOTAL INITIAL COST				
REPLACEMENT COSTS				
Year ____ @ ____ % Amount ____				
Present Worth of Future Replacement Cost				
Year ____ @ ____ % Amount ____				
Present Worth of Future Replacement Cost				
Year ____ @ ____ % Amount ____				
Present Worth of Future Replacement Cost				
ANNUAL COSTS				
Amortized Initial Cost @ ____ % ____ Year				
Capital Recovery of the Present Worth of the Replacement Cost				
(a) Year				
(b) Year				
(c) Year				
Annual Costs				
(a) Maintenance				
(b) Operations				
(c)				
TOTAL ANNUAL COSTS				
Annual Difference				
PRESENT WORTH OF ANNUAL DIFFERENCE				

Life Cycle Cost Analysis
Using Annual Owning and Operating Costs

| Project | Item | Team No. |

	Description of present and alternate designs	Present	Alt. #1	Alt. #2

Initial Costs	**Initial Costs**
	 1. Base Cost
	 2. Interface Costs
	    a.
	    b.
	    c.
	 3. Other Costs
	    a.
	    b.
	    c.
	 Total Initial Cost Impact (IC)
	 Initial Cost Savings			

Replacement Costs	**Single Expenditures**
	 1. Year @ % Amount
	    Amount X Present Worth (PW) Factor
	 2. Year @ % Amount
	    Amount X PW Factor
	 3. Year @ % Amount
	    Amount X PW Factor
	 4. Year @ % Amount
	    Amount X PW Factor
	 Salvage Value			

Annual Costs	**Annual Owning & Operating Costs**
	 1. Capital Recovery
	    IC X Amortization Factor (AF) where AF =
	    based on year life cycle at %.
	    Replacement Cost AF X PW
	      a. Year
	      b. Year
	      c. Year
	      d. Year
	 2. Annual Costs
	      a. Maintenance
	      b. Operations
	      c.
	 3. Total Annual Owning & Operating Costs
	    Sum of items 1. and 2. above
	 Annual Cost Savings
	 1. Annual Difference (AD)
	 2. Present Worth of Annual Difference
	    AD X PW Factor where PWF =			

	Amount of $1 at compound interest	Amount of an annuity of $1	Sinking fund to produce $1 in the future	Present value of $1	Present value of an annuity of $1	Annuity whose present value is $1	
1	1.100 000 0000	1.000 000 0000	1.000 000 0000	.909 090 9091	.909 090 9091	1.100 000 0000	1
2	1.210 000 0000	2.100 000 0000	.476 190 4762	.826 446 2810	1.735 537 1901	.576 190 4762	2
3	1.331 000 0000	3.310 000 0000	.302 114 8036	.751 314 8009	2.486 851 9910	.402 114 8036	3
4	1.464 100 0000	4.641 000 0000	.215 470 8037	.683 013 4554	3.169 865 4463	.315 470 8037	4
5	1.610 510 0000	6.105 100 0000	.163 797 4808	.620 921 3231	3.790 786 7694	.263 797 4808	5
6	1.771 561 0000	7.715 610 0000	.129 607 3804	.564 473 9301	4.355 260 6995	.229 607 3804	6
7	1.948 717 1000	9.487 171 0000	.105 405 4997	.513 158 1182	4.868 418 8177	.205 405 4997	7
8	2.143 588 8100	11.435 888 1000	.087 444 0176	.466 507 3802	5.334 926 1979	.187 444 0176	8
9	2.357 947 6910	13.579 476 9100	.073 640 5391	.424 097 6184	5.757 023 8163	.173 640 5391	9
10	2.593 742 4601	15.937 424 6010	.062 745 3949	.385 543 2894	6.144 567 1057	.162 745 3949	10
11	2.853 116 7061	18.531 167 0611	.053 963 1420	.350 493 8995	6.495 061 0052	.153 963 1420	11
12	3.138 428 3767	21.384 283 7672	.046 763 3151	.318 630 8177	6.813 691 8229	.146 763 3151	12
13	3.452 271 2144	24.522 712 1439	.040 778 5238	.289 664 3797	7.103 356 2026	.140 778 5238	13
14	3.797 498 3358	27.974 983 3583	.035 746 2232	.263 331 2543	7.366 687 4569	.135 746 2232	14
15	4.177 248 1694	31.772 481 6942	.031 473 7769	.239 392 0494	7.606 079 5063	.131 473 7769	15
16	4.594 972 9864	35.949 729 8636	.027 816 6207	.217 629 1358	7.823 708 6421	.127 816 6207	16
17	5.054 470 2850	40.544 702 8499	.024 664 1344	.197 844 6689	8.021 553 3110	.124 664 1344	17
18	5.559 917 3135	45.599 173 1349	.021 930 2222	.179 858 7899	8.201 412 1009	.121 930 2222	18
19	6.115 909 0448	51.159 090 4484	.019 546 8682	.163 507 9908	8.364 920 0917	.119 546 8682	19
20	6.727 499 9493	57.274 999 4933	.017 459 6248	.148 643 6280	8.513 563 7198	.117 459 6248	20
21	7.400 249 9443	64.002 499 4426	.015 624 3898	.135 130 5709	8.648 694 2907	.115 624 3898	21
22	8.140 274 9387	71.402 749 3868	.014 005 0630	.122 845 9736	8.771 540 2643	.114 005 0630	22
23	8.954 302 4326	79.543 024 3255	.012 571 8127	.111 678 1578	8.883 218 4221	.112 571 8127	23
24	9.849 732 6758	88.497 326 7581	.011 299 7764	.101 525 5980	8.984 744 0201	.111 299 7764	24
25	10.834 705 9434	98.347 059 4339	.010 168 0722	.092 295 9982	9.077 040 0182	.110 168 0722	25
26	11.918 176 5377	109.181 765 3773	.009 159 0386	.083 905 4529	9.160 945 4711	.109 159 0386	26
27	13.109 994 1915	121.099 941 9150	.008 257 6423	.076 277 6844	9.237 223 1556	.108 257 6423	27
28	14.420 993 6106	134.209 936 1065	.007 451 0132	.069 343 3495	9.306 566 5051	.107 451 0132	28
29	15.863 092 9717	148.630 929 7171	.006 728 0747	.063 039 4086	9.369 605 9137	.106 728 0747	29
30	17.449 402 2689	164.494 022 6889	.006 079 2483	.057 308 5533	9.426 914 4670	.106 079 2483	30
31	19.194 342 4958	181.943 424 9578	.005 496 2140	.052 098 6848	9.479 013 1518	.105 496 2140	31
32	21.113 776 7454	201.137 767 4535	.004 971 7167	.047 362 4407	9.526 375 5926	.104 971 7167	32
33	23.225 154 4199	222.251 544 1989	.004 499 4063	.043 056 7643	9.569 432 3569	.104 499 4063	33
34	25.547 669 8619	245.476 698 6188	.004 073 7064	.039 142 5130	9.608 574 8699	.104 073 7064	34
35	28.102 436 8481	271.024 368 4806	.003 689 7051	.035 584 1027	9.644 158 9726	.103 689 7051	35
36	30.912 680 5329	299.126 805 3287	.003 343 0638	.032 349 1843	9.676 508 1569	.103 343 0638	36
37	34.003 948 5862	330.039 485 8616	.003 029 9405	.029 408 3494	9.705 916 5063	.103 029 9405	37
38	37.404 343 4448	364.043 434 4477	.002 746 9250	.026 734 8631	9.732 651 3694	.102 746 9250	38
39	41.144 777 7893	401.447 777 8925	.002 490 9840	.024 304 4210	9.756 955 7903	.102 490 9840	39
40	45.259 255 5682	442.592 555 6818	.002 259 4144	.022 094 9282	9.779 050 7185	.102 259 4144	40
41	49.785 181 1250	487.851 811 2499	.002 049 8028	.020 086 2983	9.799 137 0168	.102 049 8028	41
42	54.763 699 2375	537.636 992 3749	.001 859 9911	.018 260 2712	9.817 397 2880	.101 859 9911	42
43	60.240 069 1612	592.400 691 6124	.001 688 0466	.016 600 2465	9.833 997 5345	.101 688 0466	43
44	66.264 076 0774	652.640 760 7737	.001 532 2365	.015 091 1332	9.849 088 6678	.101 532 2365	44
45	72.890 483 6851	718.904 836 8510	.001 391 0047	.013 719 2120	9.862 807 8798	.101 391 0047	45
46	80.179 532 0536	791.795 320 5361	.001 262 9527	.012 472 0109	9.875 279 8907	.101 262 9527	46
47	88.197 485 2590	871.974 852 5897	.001 146 8221	.011 338 1918	9.886 618 0825	.101 146 8221	47
48	97.017 233 7849	960.172 337 8487	.001 041 4797	.010 307 4470	9.896 925 5295	.101 041 4797	48
49	106.718 957 1634	1057.189 571 6336	.000 945 9041	.009 370 4064	9.906 295 9359	.100 945 9041	49
50	117.390 852 8797	1163.908 528 7970	.000 859 1740	.008 518 5513	9.914 814 4872	.100 859 1740	50
51	129.129 938 1677	1281.299 381 6766	.000 780 4577	.007 744 1375	9.922 558 6247	.100 780 4577	51
52	142.042 931 9844	1410.429 319 8443	.000 709 0040	.007 040 1250	9.929 598 7498	.100 709 0040	52
53	156.247 225 1829	1552.472 251 8287	.000 644 1339	.006 400 1137	9.935 998 8634	.100 644 1339	53
54	171.871 947 7012	1708.719 477 0116	.000 585 2851	.005 818 2851	9.941 817 1486	.100 585 2336	54
55	189.059 142 4713	1880.591 424 7128	.000 531 7476	.005 289 3501	9.947 106 4987	.100 531 7476	55
56	207.965 056 7184	2069.650 567 1841	.000 483 1734	.004 808 5001	9.951 914 9988	.100 483 1734	56
57	228.761 562 3902	2277.615 623 9025	.000 439 0556	.004 371 3637	9.956 286 3626	.100 439 0556	57
58	251.637 718 6293	2506.377 186 2927	.000 398 9822	.003 973 9670	9.960 260 3296	.100 398 9822	58
59	276.801 490 4922	2758.014 904 9220	.000 362 5796	.003 612 6973	9.963 873 0269	.100 362 5796	59
60	304.481 639 5414	3034.816 395 4142	.000 329 5092	.003 284 2703	9.967 157 2972	.100 329 5092	60

.1 *per period*

ANNUALLY
If compounded *annually* nominal annual rate is

10%

SEMIANNUALLY
If compounded *semiannually* nominal annual rate is

20%

QUARTERLY
If compounded *quarterly* nominal annual rate is

40%

MONTHLY
If compounded *monthly* nominal annual rate is

120%

ELECTRICAL COST SUMMARY

A.B.C. P.C.
Consulting Engineers

Estimate Type _____ Date of Estimate _____

Project _____

Location _____

Building Type _____ Gross Area _____

Building Description _____

Bid Date _____

Start Construction _____ Complete Construction _____

a. Direct Electrical Costs—Building			
b. General Conditions			
c. Subtotal Direct Cost			
d. Building Factor			
e. Design Contingency			
f. Building-Cost Subtotal			
g. Contractor Overhead and Profit			
h. Subtotal			
i. Escalation to			
j. Construction Contingency			
k. Building Cost Total			
l. Direct Cost—Site Work			
m. Design Contingency			
n. Contractor Overhead and Profit			
o. Subtotal—Site Work			
p. Escalation to			
q. Construction Contingency			
r. Site Work Total			
s. Construction Cost			

a. Direct Building Costs: including geographical adjustment.

b. General Conditions: a percentage of line a.

c. Subtotal Direct Cost: line a + line b.

d. Building Factor.

e. Design Contingency: a percentage of line c.

f. Building-Cost Subtotal: line c + line e.

g. Contractor Overhead and Profit: a percentage of line f.

h. Subtotal: line f + line g.

i. Escalation to Selected Date: a percentage of line h.

j. Construction Contingency: a percentage of (line h + line i).

k. Building-Cost Total: line h + line i + line j.

l. Site-Work Cost.

m. Design Contingency: a percentage of line l.

n. Contractor Overhead and Profit: a percentage of line l.

o. Site Subtotal: line l + line m + line n.

p. Escalation to Date: a percentage of line o.

q. Construction Contingency: a percentage of (line o + line p).

r. Site-Work Total: line o + line p + line q.

s. Construction Cost: line k + line r.

14.0 S.F., C.F. and % of TOTAL COSTS cont'd.

	UNIT	LOW	1/4	MEDIAN	3/4	HIGH
MEDICAL CLINICS cont'd. Electrical	%	4.3%	7.9%	9.7%	11.7%	19.4%
Total: Mechanical & Electrical	"	15.0%	25.4%	29.4%	35.2%	49.3%
MEDICAL OFFICES Total project costs	S.F.	13.80	29.20	36.55	42	60
	C.F.	.80	2.40	3.10	3.70	5.50
Plumbing	S.F.	1	2.10	2.90	4.05	7.80
Heating, ventilating, air conditioning	↓	1.15	2.50	3.75	4.85	9.50
Electrical		1	2.70	3.50	4.30	9.50
Total: Mechanical & Electrical		3.60	8.40	10.25	13.25	23.50
Percentage of total: Plumbing	%	3.2%	6.5%	8.7%	10.8%	18.3%
Heating, ventilating, air conditioning		3.7%	7.6%	10.0%	12.8%	27.8%
Electrical		3.8%	8.1%	9.7%	11.8%	24.0%
Total: Mechanical & Electrical	↓	15.2%	24.8%	29.8%	34.1%	59.1%
MOTELS Total project costs	S.F.	9.25	20.75	27.75	32.60	61
	C.F.	1.15	1.55	2.80	3.20	5.30
Plumbing	S.F.	1.75	2.20	3.15	3.60	5.35
Heating, ventilating, air conditioning	↓	1	1.50	2.35	3.75	5.00
Electrical		1.15	2.05	3.05	3.80	6.70
Total: Mechanical & Electrical		4	5.85	8.85	10.80	16.95
Percentage of total: Plumbing	%	3.9%	8.8%	10.9%	12.6%	9.5%
Heating, ventilating, air conditioning		4.0%	5.5%	8.4%	11.1%	17.8%
Electrical		4.5%	8.2%	10.1%	12.1%	21.3%
Total: Mechanical & Electrical	↓	13.9%	25.1%	29.4%	34.1%	47.0%
Per rental unit, total cost	Unit	4,500	8,850	12,000	17,700	29,000
Total: Mechanical & Electrical	"	1,900	2,530	3,650	5,230	10,650
NURSING HOMES Total project cost	S.F.	13.90	29.85	36	44.35	87
	C.F.	1.10	2.35	2.95	3.90	5.40
Plumbing	S.F.	1.20	2.55	3.25	4.15	6.65
Heating, ventilating, air conditioning	↓	1.40	2.90	4.35	6.60	12.35
Electrical		1.30	2.80	3.65	4.55	7.90
Total: Mechanical & Electrical		4.00	8.40	11.35	13.75	25.85
Percentage of total: Plumbing	%	3.0%	8.3%	9.5%	11.0%	18.8%
Heating, ventilating, air conditioning		5.6%	9.4%	12.6%	16.5%	26.5%
Electrical		5.4%	8.4%	10.3%	12.7%	25.9%
Total: Mechanical & Electrical	↓	15.3%	28.5%	32.8%	37.7%	62.0%
Per bed or person, total cost	Bed	3,100	9,500	12,700	19,600	70,000
Total: Mechanical & Electrical	"	950	2,670	3,960	5,750	11,500
OFFICES Total project costs	S.F.	11.90	26.50	34.95	45.65	94
	C.F.	.85	2.00	2.70	3.45	7.00
Site work	S.F.	.20	.75	1.70	3.35	17
Masonry		.27	1.00	2.55	4.15	15
Miscellaneous metals		.07	.49	.85	1.60	2.75
Water & dampproofing		.03	.04	.09	.23	.84
Roofing		.06	.22	.46	.90	4.30
Finish hardware		.03	.13	.22	.35	1.40
Windows		.03	.23	.75	1.30	1.80
Glass & glazing		.06	.43	.75	1.40	4.30
Tile & marble		.05	.16	.26	.41	2.55
Floor covering		.04	.17	.41	.61	1.80
Painting		.08	.28	.44	.75	2.70
Elevators		.36	.80	1.35	1.95	4.90
Plumbing		.41	1.15	1.80	2.65	9.80
Heating, ventilating, air conditioning		1.20	3.10	4.40	6.35	20.60
Electrical		1	2.75	3.65	4.85	19
Total: Mechanical & Electrical	↓	2.75	7.30	9.75	13.65	42.90
Percentage of total: Site work	%	0.5%	2.2%	4.9%	7.0%	14.9%
Masonry		0.8%	2.5%	4.7%	9.4%	22.6%
Miscellaneous metals		0.1%	0.9%	1.4%	2.4%	5.3%
Water & dampproofing		0.05%	0.1%	0.3%	0.4%	1.4%
Roofing		0.2%	0.6%	1.2%	1.8%	4.5%
Finish hardware	↓	0.1%	0.3%	0.5%	0.8%	2.7%

INSTRUCTIONS | Page 10 | Variable Factors

NATIONAL ELECTRICAL CONTRACTORS ASSOCIATION, INC.
MANUAL OF LABOR UNITS

DATE _____ JOB OR ESTIMATE NO. _____ JOB NAME _____

VARIABLE FACTORS	DEGREE 1	MAX. %	DEGREE 2	MAX. %	DEGREE 3	MAX. % AV.	DEGREE 4	MAX. %	DEGREE 5	MAX. %	% AS-SIGNED
TYPE OF BUILDING		−5		5		10		20		30	
1. Construction	Standard	0		1		2		4	Special	6	
2. Design	Simple	−1		1		2		4	Elaborate	6	
3. Floor Plan	Uniform	0		1		2		4	Complex	5	
4. Occupancy–Use	Usual	−1		0		1		2	Special	3	
5. Size–Floor Area	Small	−1		0		1		1	Large	2	
6. Extent of Electrical System	High Density	0		1		1		2	Low Density	3	
7. Design of Electrical System	Simple	−1		1		1		2	Complex	3	
8. Quality of Electrical Layout	Excellent	−1		0		0		1	Poor	2	
WORKING CONDITIONS		−5		3		10		20		30	
1. Location of Job	Close In	−1		0		1		2	Out of Town	3	
2. Weather Conditions	Excellent	0		1		2		4	Extremely Bad	6	
3. Working Space	Large–Clear	−1		0		1		2	Small-Cluttered	3	
4. Amt. Work by Other Trades	Very Little	−1		0		1		2	Very Congested	3	
5. Material Storage	Excellent	−1		0		1		2	None	3	
6. Shop & Bench Space	Excellent	0		1		2		4	None	6	
7. Material Hoisting Conditions	Excellent	−1		1		2		4	None	6	
8. Other											
9.											
GENERAL CONTRACTOR		−5		5		10		20		30	
1. Experience with Work	Considerable	−1		1		2		4	None	6	
2. Progress Maintained	Excellent	−2		2		4		8	Poor	12	
3. Coordination of Trades	Excellent	−2		2		4		8	Poor	12	
4. Other											
5.											
ELECTRICAL CONTRACTOR		−5		5		10		20		30	
1. Experience with Work	Considerable	−1		1		2		4	None	6	
2. Experienced Supervision	Available	−2		2		4		8	None	12	
3. Experienced Workers	100% Available	−1		1		2		4	None	6	
4. Adequate Tools & Equip.	Adequate	−1		1		2		4	None	6	
5. Other											

Add 1% For Each of _____ Floors

Total Job Factor _____ %

APPENDIX 181

BIBLIOGRAPHY

1. William R. Park, *Cost Engineering Analysis*. John Wiley & Sons, New York, 1973.
2. William Dudley Hunt, Jr., ed., *Creative Control of Building Costs*. McGraw-Hill Book Co., New York, 1967.
3. American Institute of Architects, Standard Form of Agreement between Owner and Architect. A.I.A. Document B141, 1974.
4. ASHRAE, *Handbook of Fundamentals*, 1972.
5. *Trane Air Conditioning Manual*.
6. *New York Times*, March 13, 1976.
7. *Value Engineering Workbook for Construction Grant Projects*. Municipal Construction Division, Office of Water Program Operation, Enviromental Protection Agency, July 1976.
8. Alphonse Dell'Isola, *Value Engineering in the Construction Industry*. Construction Publishing Co., Inc., New York, 1974.
9. McKee–Berger–Mansueto, Design Cost File, 1976. Construction Publishing Co., Inc., New York, 1975.

INDEX

accuracy, estimate, 27
air conditioning and control, cost chart, 148, 149
air conditioning motors, load chart, 147
aluminum feeders 3 phase 3 wire, cost chart, 134, 135
annual costs, 60
annuity 59

basic building, 88
bid
    adjustment factors, 15
    climate, 41
    strategy, 14
    timing of, 15
branch circuit, 94
    number of, chart, 144, 145
    wiring cost chart, 140, 142, 143
budget estimate, 49
bus duct, cost chart, 136

capital recovery, 59
cardiac care unit, 113
change orders, 21
chart method of estimating, 22
charts, use of, 115
clock systems, 107
construction, period of, 6, 7
contingencies, 8, 36
contractor type estimate, 21, 24
conversion chart, load, 118
copper feeders, charts, 129–132
cost
    consultant, 191, 31
    control, 32, 41
    estimate, 24
    model, 41, 51
    per square foot, 17, 90
    summary, 155
cost-capacity method, 17, 49
C P M, 7
cutting and patching, 5

demolition, 5
distribution equipment, 97
dry type transformers, cost chart, 126

efficiency, 85
emergency
    generator, cost chart, 137
    power and light, 100
energy, temporary elect, system, 4
escalation, 10
    example, 10, 13
    index, 12
estimate
    accuracy of, 27
    budget, 49
    chart method, 22
    contractor type, 21, 24
    fifty percent, 27, 50
    final, 50
    form, 153
    outline, 154
    part plan, 21
    pre bid, 27
    pre design, 26
    preliminary, 50
    program, 48
    quarter final, 27
    schematic, 27, 49

factors, variable, contractor, 5
feeder bus duct, cost chart, 136
feeders
    aluminum, cost chart, 134, 135
    copper, cost chart, 129–132
    light and power, cost chart, 133
    lighting, 99
    motors, 99
        cost chart, 128–135
fifty percent estimate, 27, 50
final estimate, 50
fire alarm systems, 109

general conditions, 3
    check list, 2
generator, emergency, 137

high voltage work, 96

illumination, 103
insurance, liability, 3

intensive care unit, 113
intercommunication systems, 104

kitchen, 113

labor intensity factor, 70
liability insurance, 3
L.I.F., 70
life cycle analysis, 55, 66
    cost, 55
light and power feeders, cost chart, 133
light fixture charts, 151, 152
lighting and receptacle panels, cost
    chart, 124, 125
lighting feeders, 99
lighting outlet box, 90
    cost chart, 138
lightning protection, 110
light switch, 93

main distribution panel, cost chart, 123
manning curve, 7
manpower requirements, 82, 83
market conditions, 14
material cost index, 70
M.C.I., 70
motor
    air conditioning, 147–149
    and controller, 95
    connection, cost chart, 146, 148, 149
    control center, 96
    feeders, 99
        cost chart, 128

nurse call system, 108

one line diagram, 88
operating room, 113
other systems, 102
overhead, 9
overtime, effect, 85

panels
    light and power, 99
    lighting, cost chart, 127
    main distribution, cost chart, 123
    power, cost chart, 123

part plan estimate, 21
payment schedule, 86, 87
performance bond, 3
power transformer, cost chart, 120
pre bid estimate, 26
pre design estimate, 26
preliminary estimate, 50
present worth, 56
primary
    service switch, cost chart, 119
    wire and conduit, cost chart, 117
productivity, 68
profit, 9
program estimate, 48
project manager, 5

quarter final estimate, 27

receptacle, 92
    branch circuit, cost chart, 142
    load chart, 141

schedule, accelerated, 7
schematic estimate, 27, 49
secondary service and metering, cost
    chart, 121, 122
service
    primary, cost chart, 119
    secondary, cost chart, 121
sound systems, 106
switch and receptacle, cost chart, 139

T.C.F., 71
television, 107
temporary facilities, 4
temporary light and power, 112
total cost factor, 71
transformer
    dry type, cost chart, 126
    power, cost chart, 120

value engineering, 33, 43, 53

wage rate, 67
    effective, 70
watchman, 3